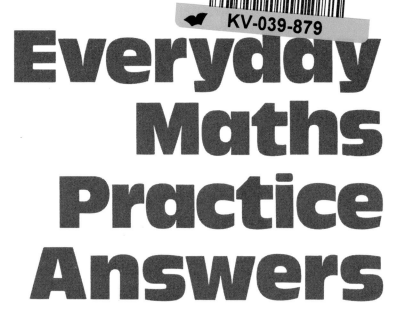

Everyday Maths Practice Answers

R.Christon and P.Newton

3rd edition

Oxford University Press

Oxford University Press, Walton Street, Oxford OX2 6DP

Oxford New York Toronto
Delhi Bombay Calcutta Madras Karachi
Petaling Jaya Singapore Hong Kong Tokyo
Nairobi Dar es Salaam Cape Town
Melbourne Auckland

and associated companies in

Berlin Ibadan

Oxford is a trade mark of Oxford University Press

ISBN 0 19 914290 4
© Oxford University Press

First published 1982
Second edition 1986
Third edition 1988
Reprinted with corrections 1991

Printed in Great Britain by St Edmundsbury Press

Part 1: Using Number

Section 1
Number recognition
Page 4

Exercise 1

1a. 5	b. 11	c. 17
2a. 36	b. 100	c. 108
3a. 214	b. 467	
4a. 880	b. 1000	
5a. 1003	b. 1092	
6a. 2100	b. 2304	
7a. 3616	b. 5000	
8a. 5803	b. 7930	
9a. 9909	b. 10 000	
10a. 10 010	b. 10 100	

Exercise 2

1a. six b. fourteen c. twenty d. thirty-four
2a. seventy-three b. one hundred and three
c. one hundred and thirty d. one hundred and ninety-five
3a. four hundred and sixteen b. one thousand
c. one thousand two hundred and thirty-five
d. one thousand three hundred and eight
4a. two thousand six hundred and four
b. three thousand and twelve
c. three thousand five hundred and ninety-nine
d. five thousand four hundred
5a. six thousand two hundred and twenty
b. six thousand and fifty-eight
c. eight thousand nine hundred and ninety-nine
d. ten thousand

Section 2
Place value
Page 5

Exercise 1

1a. twenty	b. two hundred	c. two	d. twenty
2a. one hundred	b. one	c. one thousand	d. ten
3a. sixty	b. six hundred	c. six hundred	d. sixty
4a. nine hundred	b. nine thousand	c. ninety	d. nine
5a. five	b. five thousand	c. eight	d. eight thousand

Exercise 2

1a. seven	b. forty	c. one hundred	d. eight hundred
2a. seventy	b. six hundred	c. eight thousand	d. eighty
3a. one	b. four hundred	c. three hundred	d. sixty
4a. nine thousand	b. ten	c. twenty	d. forty
5a. one thousand	b. eight	c. eight thousand	d. thirty
6a. ten	b. six hundred	c. two thousand	d. six thousand

Section 3
Addition
Page 6

Exercise 1

1a. 9	**b.** 55	**c.** 168	**d.** 4769
2a. 8	**b.** 97	**c.** 878	**d.** 4999
3a. 14	**b.** 81	**c.** 184	**d.** 4057
4a. 18	**b.** 85	**c.** 483	**d.** 4598
5a. 127	**b.** 244	**c.** 4803	**d.** 4444
6a. 125	**b.** 611	**c.** 6874	**d.** 4092
7a. 1304	**b.** 8645	**c.** 4233	**d.** 10 224
8a. 64 758	**b.** 72 288	**c.** 74 358	**d.** 66 882

9. 18 324 **10.** 20 926 metres **11.** 13 847 **12.** 37 639
13. 77 422 **14.** £135 544 **15.** 108 523

Section 4
Subtraction
Page 7

Exercise 1

1a. 5	**b.** 11	**c.** 12	**d.** 35
2a. 133	**b.** 122	**c.** 2322	**d.** 3123
3a. 7	**b.** 6	**c.** 4	**d.** 37
4a. 118	**b.** 216	**c.** 219	**d.** 208
5a. 1173	**b.** 6273	**c.** 1194	**d.** 2184
6a. 47	**b.** 249	**c.** 255	**d.** 169
7a. 2926	**b.** 3625	**c.** 2645	**d.** 4829
8a. 1928	**b.** 1975	**c.** 2889	**d.** 2767
9a. 2774	**b.** 3547	**c.** 3277	**d.** 138
10a. 2362	**b.** 2131	**c.** 1164	**d.** 3729
11a. 18 113	**b.** 9923	**c.** 29 392	**d.** 24 869

12. 56 miles **13.** 37 **14.** 476 **15.** 489 **16.** Labour by 8299
17. 36 314

Section 5
**Multiplication
by 10, 100
and 1000**
Page 8

Exercise 1

1a. 20	**b.** 400	**c.** 200	**d.** 40
2a. 160	**b.** 1200	**c.** 16 000	**d.** 120
3a. 630	**b.** 5400	**c.** 82 000	**d.** 9410
4a. 120	**b.** 1800	**c.** 8100	**d.** 4000
5a. 100	**b.** 10 000	**c.** 1000	**d.** 10 000

6. 50 mins **7a.** 40p **b.** 100p or £1
8a. £2 **b.** £1.50 **9a.** £250 **b.** £2500
10. 200

Section 6
**Division by
10, 100 and
1000**
Page 9

Exercise 1

1a. 2	**b.** 4	**c.** 200	**d.** 20
2a. 16	**b.** 14	**c.** 12	**d.** 140
3a. 600	**b.** 6	**c.** 60	**d.** 81
4a. 100	**b.** 10	**c.** 1	**d.** 1
5a. 6	**b.** 12	**6a.** 8	**b.** 80
7a. 10	**b.** 12 miles	**8a.** 80 metres	**b.** 100 metres

Section 7
**Multiplication
Tables**
Page 10

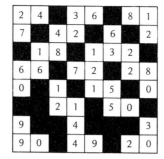

Section 8
**Multiplication
by numbers
less than 10**
Page 11

Exercise 1

1a. 54   **b.** 56   **c.** 96   **d.** 90
2a. 39   **b.** 88   **c.** 62   **d.** 99
3a. 48   **b.** 120   **c.** 252   **d.** 405
4a. 488   **b.** 693   **c.** 936   **d.** 226
5a. 378   **b.** 612   **c.** 798   **d.** 387
6. £2160   **7.** £1980   **8.** 4600 doubloons   **9.** £7605   **10.** £27 000
11. £26 664   **12.** £134 100   **13.** £5850

Section 9
**Division by
numbers less
than 10**
Page 12

Exercise 1

1a. 13   **b.** 27   **c.** 123   **d.** 93   **e.** 156
2a. 12   **b.** 14   **c.** 231   **d.** 83   **e.** 144
3a. 21   **b.** 14   **c.** 122   **d.** 71   **e.** 137
4a. 11   **b.** 15   **c.** 101   **d.** 71   **e.** 53
5a. 10   **b.** 14   **c.** 110   **d.** 61   **e.** 72
6a. 11   **b.** 13   **c.** 100   **d.** 71   **e.** 75
7a. 10   **b.** 12   **c.** 101   **d.** 81   **e.** 86
8a. 11   **b.** 7   **c.** 111   **d.** 81   **e.** 85
9a. 13 r 1   **b.** 39 r 1   **c.** 142 r 1   **d.** 218 r 1   **e.** 256 r 1
10a. 21 r 2   **b.** 15 r 2   **c.** 211 r 1   **d.** 117 r 2   **e.** 145 r 1
11a. 11 r 3   **b.** 14 r 1   **c.** 212 r 1   **d.** 114 r 1   **e.** 134 r 3
12a. 11 r 4   **b.** 15 r 1   **c.** 111 r 3   **d.** 113 r 3   **e.** 136 r 4
13a. 11 r 2   **b.** 12 r 2   **c.** 113 r 3   **d.** 114 r 1   **e.** 123 r 1
14a. 11 r 2   **b.** 12 r 1   **c.** 111 r 1   **d.** 112 r 2   **e.** 119 r 3
15a. 11 r 1   **b.** 12 r 1   **c.** 111 r 1   **d.** 111 r 7   **e.** 116 r 7
16a. 10 r 5   **b.** 30 r 5   **c.** 100 r 3   **d.** 104 r 3   **e.** 58 r 4
17. £14   **18.** 424   **19.** 371   **20.** £527

Section 10
**Long
multiplication**
Page 13

Exercise 1

1a. 4056   **b.** 9020   **c.** 3350   **d.** 7881
2a. 21 216   **b.** 24 034   **c.** 13 984   **d.** 25 192
3a. 35 342   **b.** 17 847   **c.** 24 476   **d.** 12 949
4a. 34 529   **b.** 17 639   **c.** 41 078   **d.** 14 198
5a. 49 848   **b.** 14 076   **c.** 10 478   **d.** 14 496
6a. 43 830   **b.** 36 708   **c.** 107 576   **d.** 449 625
7. £104 175   **8.** £219 800   **9.** 116 688   **10.** 114 000
11. £14 875   **12.** £5940   **13.** 168 480   **14.** 20 736   **15.** 100 750

Section 11
Factors
Page 14

Exercise 1

1a. {1, 2, 4, 8}   **b.** {1, 2, 7, 14}
 c. {1, 2, 4, 5, 10, 20}   **d.** {1, 2, 3, 4, 6, 8, 12, 24}

2a. {1, 2, 3, 5, 6, 10, 15, 30}   **b.** {1, 2, 4, 8, 16, 32}
 c. {1, 2, 3, 4, 6, 9, 12, 18, 36}   **d.** {1, 2, 4, 11, 22, 44}

3a. {1, 2, 4, 13, 26, 52}   **b.** {1, 2, 4, 7, 8, 14, 28, 56}
 c. {1, 2, 4, 8, 16, 32, 64}   **d.** {1, 2, 3, 4, 6, 8, 9, 12, 18, 36, 72}

Exercise 2

1a. {2, 3}   **b.** {2}   **c.** {2, 3}   **d.** {3, 5}

2a. {3, 7}   **b.** {2, 11}   **c.** {2, 13}   **d.** {2, 3, 5}

3a. {3, 11}   **b.** {5, 7}   **c.** {2, 3, 7}   **d.** {2, 3}

Exercise 3

1a. $2 \times 2 \times 2$ **b.** $2 \times 2 \times 3$ **c.** 2×7 **d.** $2 \times 3 \times 3$

2a. 3×7 **b.** $2 \times 2 \times 2 \times 3$ **c.** $2 \times 3 \times 5$ **d.** $2 \times 2 \times 3 \times 3$

3a. $2 \times 3 \times 7$ **b.** $3 \times 3 \times 5$ **c.** 3×17 **d.** $2 \times 3 \times 3 \times 3$

Section 12
**Estimation and
approximation**
Page 15

Exercise 1

1a. 10   **b.** 20   **c.** 20   **d.** 40
2a. 80   **b.** 90   **c.** 90   **d.** 100
3a. 160   **b.** 250   **c.** 360   **d.** 480

Exercise 2

1a. 100   **b.** 100   **c.** 200   **d.** 200
2a. 200   **b.** 300   **c.** 300   **d.** 300
3a. 800   **b.** 900   **c.** 900   **d.** 1000

Exercise 3

1a. 1000   **b.** 1000   **c.** 1000   **d.** 2000
2a. 5000   **b.** 6000   **c.** 6000   **d.** 7000
3a. 9000   **b.** 9000   **c.** 9000   **d.** 10000

Exercise 4

1a. 80   **b.** 120
2a. 140   **b.** 390
3a. 20   **b.** 40
4a. 70   **b.** 330
5a. 400   **b.** 1200
6a. 26 (or 25) **b.** 4

Section 13
Number patterns and sequences
Page 16

Exercise 1

1. 1, 3, 15
2. 3, 6, 15
3. 10, 15
4. 3
5. 4
6. 5
7. The answers increase by 1
8. 21
9. 28
10.

Exercise 2

1.	9, 11	**2.**	10, 12
3.	16, 32	**4.**	12, 16
5.	50, 60	**6.**	35, 42
7.	54, 63	**8.**	96, 108
9.	16, 25	**10.**	15, 21
11.	8, 4	**12.**	3, 1
13.	40, 20	**14.**	35, 30
15.	19, 15	**16.**	22, 15

Section 14
Directed numbers
Page 17

Exercise 1

1. Newquay
2. Scarborough
3. Newquay
4. 5°C
5. 7°C
6. 6°C
7. Folkstone 9°C
8. 6°C and 1°C
9. 7°C and −1°C
10. 6
11. Scarborough
12. Newquay

Section 15
Direct proportion I
Page 18

Exercise 1

1. £16·20
2. £1·08
3. 64p
4. £1·92
5. £14·85
6. 95p
7. £7·65
8. Find cost of 1000 g for each.
   a. 147p   b. 150p   c. 144p
   The Great Dane pack is the best buy.
9. £8.40
10. £10.45

Section 16
Direct proportion II
Page 19

Exercise 1

1. 400 g prawns, 4 tablespoons sherry, 1 teaspoon salt, 4 teaspoons oil, 2 tablespoons ginger root
2. 100 g prawns, 1 tablespoon sherry, ¼ teaspoon salt, 1 teaspoon oil, ½ tablespoon ginger root
3. 300 g, 3 tablespoons sherry, ¾ teaspoon salt, 3 teaspoons oil, 1½ tablespoons ginger root
4. £2·50
5. £1·25
6. £3·75

Exercise 2

1. 17½ m^2
2. 875 g
3. 5 m^2
4. 4
5. 57 g

Section 17
Fractions
Page 20

Exercise 1

1. $6, 1, \frac{1}{6}, \frac{5}{6}$
2. $8, 3, \frac{3}{8}, \frac{5}{8}$
3. $8, 5, \frac{5}{8}, \frac{3}{8}$
4. $2, 1, \frac{1}{2}, \frac{1}{2}$
5. $4, 2, \frac{2}{4}, \frac{2}{4}$
6. $8, 7, \frac{7}{8}, \frac{1}{8}$
7. $2, 2, 1, 0$
8. $4, 3, \frac{3}{4}, \frac{1}{4}$
9. $8, 6, \frac{6}{8}, \frac{2}{8}$
10. $16, 12, \frac{12}{16}, \frac{4}{16}$
11. $5, 3, \frac{3}{5}, \frac{2}{5}$
12. $8, 5, \frac{5}{8}, \frac{3}{8}$
13. $8, 3, \frac{3}{8}, \frac{5}{8}$
14. $8, 3, \frac{3}{8}, \frac{5}{8}$
15. $12, 6, \frac{6}{12}, \frac{6}{12}$

Section 18
Shading in fractions
Page 22

Exercise 1

1.   2.   3.   4. 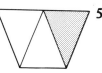 5.

6.   7.   8.   9.   10.

11.   12.   13.   14. 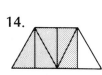 15.

Section 19
Equivalent fractions
Page 24

Exercise 1

1. $\frac{1}{4} = \frac{2}{8} = \frac{4}{16}$ 2. $\frac{1}{2} = \frac{2}{4} = \frac{4}{8}$ 3. $\frac{1}{5} = \frac{2}{10}$

4. $\frac{2}{3} = \frac{4}{6} = \frac{12}{18}$ 5. $\frac{1}{2} = \frac{2}{4} = \frac{4}{8}$ 6. $\frac{1}{5} = \frac{2}{10} = \frac{20}{100}$

Section 20
Finding equivalent fractions
Page 25

Exercise 1

1a. $\frac{2}{4}, \frac{3}{6}, \frac{4}{8}$ b. $\frac{2}{6}, \frac{3}{9}, \frac{4}{12}$ c. $\frac{2}{8}, \frac{3}{12}, \frac{4}{16}$ d. $\frac{2}{10}, \frac{3}{15}, \frac{4}{20}$ e. $\frac{2}{12}, \frac{3}{18}, \frac{4}{24}$

2a. $\frac{2}{16}, \frac{3}{24}, \frac{4}{32}$ b. $\frac{2}{20}, \frac{3}{30}, \frac{4}{40}$ c. $\frac{4}{6}, \frac{6}{9}, \frac{8}{12}$ d. $\frac{6}{8}, \frac{9}{12}, \frac{12}{16}$ e. $\frac{4}{10}, \frac{6}{15}, \frac{8}{20}$

3a. $\frac{6}{10}, \frac{9}{15}, \frac{12}{20}$ b. $\frac{8}{10}, \frac{12}{15}, \frac{16}{20}$ c. $\frac{10}{12}, \frac{15}{18}, \frac{20}{24}$ d. $\frac{6}{16}, \frac{9}{24}, \frac{12}{32}$ e. $\frac{14}{16}, \frac{21}{24}, \frac{28}{32}$

4a. $\frac{6}{20}, \frac{9}{30}, \frac{12}{40}$ b. $\frac{14}{20}, \frac{21}{30}, \frac{28}{40}$ c. $\frac{18}{20}, \frac{27}{30}, \frac{36}{40}$ d. $\frac{10}{24}, \frac{15}{36}, \frac{20}{48}$ e. $\frac{22}{24}, \frac{33}{36}, \frac{44}{48}$

Section 21
Completing equivalent fractions
Page 26

Exercise 1

1a. 2 b. 4 c. 4 d. 21
2a. 4 b. 8 c. 18 d. 6
3a. 16 b. 16 c. 15 d. 6
4a. 30 b. 18 c. 20 d. 9
5a. 200 b. 150 c. 6 d. 500
6a. 5 b. 10 c. 6 d. 1
7a. 3 b. 5 c. 10 d. 4
8a. 6 b. 4 c. 54 d. 45
9a. 6 b. 20 c. 3 d. 8
10a. 24 b. 25 c. 3 d. 5
11a. 2 b. 30 c. 5 d. 11
12a. 2 b. 7 c. 1 d. 12
13a. 2 b. 9 c. 2 d. 7
14a. 1 b. 10 c. 12 d. 15
15a. 13 b. 1 c. 2 d. 1
16a. 14 b. 15 c. 9 d. 3

Section 22
Cancelling down equivalent fractions
Page 27

Exercise 1

1a. 2	b. 2	c. 4	d. 3
2a. 3	b. 5	c. 5	d. 10
3a. 4	b. 6	c. 7	d. 8
4a. 6	b. 4	c. 5	d. 10
5a. 10	b. 50	c. 100	d. 1

Exercise 2

1a. $\frac{1}{2}$	b. $\frac{1}{2}$	c. $\frac{1}{2}$	d. $\frac{1}{2}$
2a. $\frac{1}{2}$	b. $\frac{1}{3}$	c. $\frac{3}{4}$	d. $\frac{1}{3}$
3a. $\frac{2}{3}$	b. $\frac{2}{3}$	c. $\frac{1}{10}$	d. $\frac{1}{2}$
4a. $\frac{3}{5}$	b. $\frac{4}{7}$	c. $\frac{5}{6}$	d. $\frac{7}{8}$
5a. $\frac{3}{4}$	b. $\frac{2}{5}$	c. $\frac{1}{10}$	d. $\frac{1}{2}$
6a. $\frac{2}{5}$	b. $\frac{3}{4}$	c. $\frac{3}{7}$	d. $\frac{2}{3}$
7a. $\frac{3}{5}$	b. $\frac{4}{7}$	c. $\frac{1}{2}$	d. $\frac{3}{5}$
8a. $\frac{3}{5}$	b. $\frac{4}{9}$	c. $\frac{2}{5}$	d. $\frac{2}{3}$

Section 23
Mixed numbers
Page 28

Exercise 1

1. $1\frac{1}{2}$, $2\frac{1}{4}$, etc.   2. 1, 2, 3, etc.
3. $\frac{8}{5}$, $\frac{6}{5}$, $\frac{11}{4}$, etc.   4. $\frac{1}{2}$, $\frac{1}{4}$, $\frac{2}{5}$, etc.

Exercise 2

1a. $1\frac{1}{3}$	b. $1\frac{2}{3}$	c. $1\frac{3}{8}$	d. $1\frac{7}{8}$	e. $2\frac{1}{6}$
2a. $1\frac{1}{6}$	b. $1\frac{5}{6}$	c. $1\frac{2}{5}$	d. $1\frac{3}{10}$	e. $2\frac{4}{5}$
3a. $4\frac{1}{2}$	b. $2\frac{1}{3}$	c. $2\frac{2}{3}$	d. $2\frac{3}{4}$	e. $5\frac{2}{3}$
4a. $5\frac{1}{2}$	b. $3\frac{2}{3}$	c. $3\frac{1}{4}$	d. $3\frac{1}{3}$	e. $1\frac{3}{5}$
5a. $2\frac{2}{5}$	b. $2\frac{3}{10}$	c. $2\frac{3}{5}$	d. $2\frac{1}{2}$	e. $9\frac{1}{2}$

Exercise 3

1a. $1\frac{1}{2}$	b. $1\frac{1}{2}$	c. $1\frac{1}{2}$	d. $2\frac{1}{5}$
2a. $1\frac{3}{4}$	b. $2\frac{1}{2}$	c. $1\frac{1}{3}$	d. $1\frac{1}{2}$
3a. $1\frac{2}{3}$	b. $2\frac{1}{2}$	c. $1\frac{1}{2}$	d. $2\frac{1}{4}$

Section 24
Using mixed numbers
Page 29

Exercise 1

1a. $4\frac{1}{2}$	b. $2\frac{1}{4}$	c. $4\frac{3}{4}$	d. $2\frac{7}{10}$
2a. $3\frac{1}{4}$	b. $2\frac{3}{5}$	c. $3\frac{3}{4}$	d. $1\frac{1}{2}$
3a. $5\frac{1}{5}$	b. $3\frac{7}{8}$	c. $3\frac{3}{8}$	d. $6\frac{1}{3}$
4a. $5\frac{2}{3}$	b. 7	c. 9	d. $4\frac{1}{2}$
5a. $5\frac{3}{10}$	b. $7\frac{3}{5}$	c. $3\frac{9}{10}$	d. $13\frac{1}{10}$

Exercise 2

1a. $\frac{7}{4}$ b. $\frac{24}{5}$ c. $\frac{13}{8}$ d. $\frac{19}{4}$

2a. $\frac{7}{2}$ b. $\frac{2}{3}$ c. $\frac{1}{2}$ d. $\frac{3}{2}$

3a. $\frac{5}{2}$ b. $\frac{7}{3}$ c. $\frac{9}{4}$ d. $\frac{11}{5}$

4a. $\frac{3}{7}$ b. $\frac{4}{5}$ c. $\frac{9}{10}$ d. $\frac{3}{4}$

5a. $\frac{4}{9}$ b. $\frac{11}{3}$ c. $\frac{13}{5}$ d. $\frac{5}{9}$

Section 25
Metric units
Page 30

Exercise 1

1a. 1 cm   b. 3 cm   c. 6 cm
2a. 1 m   b. 3 m   c. 7 m
3a. 1 km   b. 4 l m   c. 8 km
4a. 1 kg   b. 5 kg   c. 9 kg
5a. 1 t   b. 6 t   c. 8 t
6a. 1 l b. 4 l c. 7 l
7a. 60 mm   b. 120 mm   c. 200 mm
8a. 2 m   b. 5 m   c. 9 m
9a. 3 km   b. 5 km   c. 9 km
10a. 2 kg   b. 3 kg   c. 7 kg

Exercise 2

1. 10 km   2. 2 km   3. 250   4. 2000 ml
5a. 100 mm   b. 8 m

Section 26
Metric conversion
Page 31

Exercise 1

1a. 22 pounds   b.   17·6 pounds
2a. 9·9 pounds   b.   7·7 pounds
3a. 1·1 pounds   b.   0·22 pounds
4a. 0·5 kg   b.   6·4 kg
5a. 15 cm   b.   12·5 cm
6a. 11·3 cm   b.   28·8 cm
7a. 30 cm   b.   90 cm
8a. 0·4 inches   b.   0·8 inches
9a. 45 l b.   47·3 l
10a. 0·2 gallons   b.   1·11 gallons

Exercise 2

1. —
2. 11 stone
3. 76 kg
4. 10 stone 4 pounds
5. —

Section 27
Missing numbers
Page 32

Exercise 1

1a. 3	**b.** 7	**c.** 7	**d.** 91
2a. 3	**b.** 11	**c.** 15	**d.** 12
3a. 5	**b.** 5	**c.** 1	**d.** 24
4a. 7	**b.** 8	**c.** 6	

Exercise 2

1. 7
2. 5½
3. 5
4. 3
5. 0

Exercise 3

1a. 5 **2a.** 13
b. 36 **b.** 72
c. $8 \times 6 = 49 - 1$ **c.** $(15 \times 6) + 7 = 98 - 1$

Section 28
Decimal numbers
Page 33

Exercise 1

whole number	tenths	hundredths	thousandths
18	$\frac{2}{10}$	$\frac{1}{100}$	$\frac{3}{1000}$
7	$\frac{4}{10}$	$\frac{0}{100}$	$\frac{0}{1000}$
6	$\frac{8}{10}$	$\frac{1}{100}$	$\frac{1}{1000}$
1892	$\frac{0}{10}$	$\frac{0}{100}$	$\frac{2}{1000}$

Exercise 2

1000	100	10	1	$\frac{1}{10}$	$\frac{1}{100}$	$\frac{1}{1000}$
		3	4	8	3	6
			3	8	0	6
		3	9	9	9	9
			0	5	8	3
	3	4	5	7	6	
	4	9	8	3		
		3	1	0	0	7
1	5	6	7	9	2	
1	2	8	6	4	4	
2	3	1	6	8	0	4

Section 29
Using decimal numbers
Page 34

Exercise 1

1a. four point two **b.** eighteen point six
c. two point four, two **d.** one point six, one
2a. ten point eight, eight **b.** thirty-nine point zero, one
c. fifty point two, zero, two
d. one hundred and forty-eight point three
3a. two point three, nine, nine **b.** forty-five point eight, nine, zero
c. two thousand, three hundred and forty-eight point one
d. three point five, four, nine

Exercise 2

1. 42·6 **2.** 18·712 **3.** 412·93 **4.** 9·805 **5.** 57·017
6. 1006·62 **7.** 0·444 **8.** 10·909 **9.** 804·068 **10.** 6334·01

Section 30
Comparing decimal numbers
Page 34

Exercise 1

1a. 3·4, 2·8   b. 9·21, 7·49
2a. 2·1, 1·4   b. 14·99, 12·999
3a. 77·1, 17·6   b. 20·906, 19·999
4a. 4·3, 3·2, 2·3   b. 3·11, 2·44, 1·232
5a. 144·6, 74·22, 14·6   b. 129·5, 29·87, 4·736

Exercise 2

1a. 3·8, 3·6   b. 14·6, 14·2
2a. 4·70, 4·61   b. 73·8, 73·6
3a. 38·4, 38·29   b. 3·911, 3·901
4a. 17·62, 17·606   b. 512·861, 512·66
5a. 112·63, 112·333   b. 90·1, 90·009
6a. 4·6, 4·3, 4·2   b. 10·19, 10·18, 10·09
7a. 7·46, 7·44, 7·43   b. 800·9, 800·87, 800·85
8a. **62·18, 62·1**   b. 16·859, 16·85
9a. 3·446, 3·443, 3·442   b. 4·831, 4·83, 4·829
10a. 10·645, 10·644, 10·641   b. 1·726, 1·725, 1·723
11a. 3·86, 3·858, 3·856   b. 9·816, 9·815, 9·814
12a. 22·624, 22·623, 22·621   b. 91·066, 91·063, 91·06
13a. 5·341, 5·337, 5·332   b. 10·41, 10·409, 10·40
14a. 12·61, 12·60 or 12·6   b. 8·324, 8·323, 8·32
15a. 4·666, 4·66, 4·6   b. 3·91, 3·906, 3·903

Section 31
Changing decimal numbers to fractions
Page 36

Exercise 1

1a. $\frac{1}{10}$ b. $\frac{3}{10}$ c. $\frac{7}{10}$ d. $\frac{9}{10}$

2a. $\frac{1}{5}$ b. $\frac{2}{5}$ c. $\frac{3}{5}$ d. $\frac{4}{5}$

3a. $\frac{11}{100}$ b. $\frac{17}{100}$ c. $\frac{23}{100}$ d. $\frac{29}{100}$

4a. $\frac{7}{50}$ b. $\frac{13}{50}$ c. $\frac{17}{50}$ d. $\frac{23}{50}$

5a. $\frac{3}{100}$ b. $\frac{7}{100}$ c. $\frac{3}{50}$ d. $\frac{2}{25}$

6a. $\frac{7}{1000}$ b. $\frac{9}{1000}$ c. $\frac{1}{250}$ d. $\frac{1}{125}$

7a. $\frac{4}{25}$ b. $\frac{8}{25}$ c. $\frac{11}{20}$ d. $\frac{17}{20}$

8a. $\frac{13}{1000}$ b. $\frac{17}{1000}$ c. $\frac{3}{125}$ d. $\frac{9}{250}$

9a. $\frac{121}{500}$ b. $\frac{117}{250}$ c. $\frac{1}{4}$ d. $\frac{3}{4}$

10a. $\frac{1}{8}$ b. $\frac{3}{8}$ c. $\frac{5}{8}$ d. $\frac{7}{8}$

Section 32
**Changing
fractions to
decimal
numbers**
Page 36

Exercise 1

1a. 0·1   b. 0·3   c. 0·5   d. 0·8
2a. 0·01   b. 0·04   c. 0·07   d. 0·09
3a. 0·14   b. 0·36   c. 0·45   d. 0·69
4a. 0·001   b. 0·008   c. 0·011   d. 0·024
5a. 0·112   b. 0·235   c. 0·486   d. 0·888

Exercise 2

1a. 0·2   b. 0·4   c. 0·15   d. 0·06   e. 0·8
2a. 0·35   b. 0·14   c. 0·22   d. 0·55   e. 0·42
3a. 0·2   b. 0·98   c. 0·25   d. 0·5   e. 0·75
4a. 0·75   b. 0·202   c. 0·4   d. 0·406   e. 0·816
5a. 0·2   b. 0·4   c. 0·24   d. 0·52   e. 0·84

Exercise 3

1a. 1·1   b. 2·3   c. 3·6   d. 4·9   e. 5·03
2a. 6·27   b. 7·23   c. 8·44   d. 1·004   e. 2·009
3a. 3·018   b. 4·058   c. 5·124   d. 6·304   e. 7·466
4a. 8·875   b. 8·2   c. 3·4   d. 4·6   e. 6·8
5a. 2·15   b. 5·35   c. 6·65   d. 3·8   e. 4·7
6a. 1·52   b. 7·76   c. 6·82   d. 3·62   e. 8·94
7a. 7·605   b. 5·83   c. 4·63   d. 5·825   e. 4·935
8a. 4·108   b. 6·324   c. 3·492   d. 2·664   e. 7·828
9a. 8·174   b. 5·258   c. 9·424   d. 6·73   e. 5·838
10a. 2·32   b. 7·56   c. 8·84   d. 5·25   e. 6·75

Section 33
**Decimal
places**
Page 38

Exercise 1

1a. 1   b. 1   c. 1
2a. 2   b. 2   c. 2
3a. 3   b. 3   c. 3
4a. 2   b. 2   c. 2
5a. 3   b. 3   c. 3
6a. 3   b. 3   c. 3
7a. 2   b. 3   c. 2
8a. 3   b. 1   c. 3
9a. 3   b. 3   c. 3

Exercise 2

1a. 1·6   b. 8·6
2a. 4·86   b. 5·65
3a. 9·326   b. 9·983
4a. 8·64   b. 6·46
5a. 3·5   b. 8·7
6a. 4·35   b. 7·47
7a. 8·069   b. 5·404

8a. 6·29   b. 4·31
9a. 8·3   b. 6·49
10a. 12·62   b. 14·4

Section 34
**Significant
figures**
Page 39

Exercise 1

1a. 1·32   b. 8·6
2a. 12·9   b. 49·5
3a. 156·3   b. 7·27
4a. 7·3   b. 7
5a. 4·9   b. 4·9
6a. 0·73   b. 0·7
7a. 0·05   b. 0·007
8a. 4   b. 4·0
9a. 14   b. 14·9
10a. 4·0   b. 14

Section 35
**Decimals:
true or
false?**
Page 40

Exercise 1

1. T	2. F	3. T	4. T	5. F
6. T	7. F	8. F	9. T	10. T
11. T	12. F	13. F	14. T	15. T
16. T	17. F	18. F	19. T	20. F
21. F	22. T	23. F	24. F	25. T

Section 36
**Decimal
addition**
Page 41

Exercise 1

1a. 7·9   b. 6·8   c. 7·89   d. 8·837
2a. 9·2   b. 6·2   c. 6·95   d. 8·48
3a. 7·81   b. 17·92   c. 1·83   d. 1·94
4a. 29·19   b. 47·0   c. 9·22   d. 9·02
5a. 20·93   b. 49·26   c. 63·98   d. 235·88
6a. 8·2   b. 13·7   c. 52·3   d. 13·29
7a. 22·79   b. 189·734   c. 2·336   d. 14·449
8. 2·23 metres   9. 62·561

Section 37
**Decimal
subtraction**
Page 42

Exercise 1

1a. 2·2   b. 3·2   c. 3·3   d. 5·5
2a. 2·41   b. 4·22   c. 4·24   d. 2·32
3a. 3·8   b. 2·9   c. 4·6   d. 2·5
4a. 4·18   b. 3·28   c. 2·18   d. 2·48
5a. 4·92   b. 1·91   c. 1·61   d. 2·93
6a. 3·68   b. 1·88   c. 0·68   d. 3·47
7a. 2·712   b. 2·822   c. 1·813   d. 2·612
8a. 4·858   b. 2·788   c. 3·676   d. 2·868
9a. 3·28   b. 4·23   c. 3·48   d. 2·33
10a. 2·72   b. 2·84   c. 2·81   d. 4·23

Exercise 2

1a. 2·18	b. 2·25	c. 4·14	d. 3·11
2a. 3·77	b. 2·76	c. 2·58	d. 1·67
3a. 5·111	b. 3·138	c. 3·325	d. 5·224
4a. 3·364	b. 5·364	c. 2·282	d. 3·175
5a. 5·888	b. 5·887	c. 0·889	d. 2·652
6a. 56·15	b. 49·34	c. 27·27	d. 27·46
7a. 28·88	b. 38·79	c. 17·66	d. 17·87
8a. 382·37	b. 262·36	c. 465·13	d. 791·27
9a. 522·87	b. 431·63	c. 610·47	d. 732·71
10a. 585·67	b. 788·63	c. 667·89	d. 586·76

Section 38
Multiplication of decimals by numbers less than 10
Page 44

Exercise 1

1a. 8·2	b. 9·6	c. 16·2	d. 11·1
2a. 22·4	b. 49·6	c. 51·1	d. 516·65
3a. 103·6	b. 123·5	c. 37·5	d. 147·6
4a. 15·3	b. 50·72	c. 33·81	d. 37·28
5a. 147·0	b. 42·91	c. 140·05	d. 56·16
6a. 10·7	b. 7·22	c. 12·27	
7a. 9·18	b. 20·16	c. 16·83	
8a. 91·8	b. 19·52	c. 36·2	

9. 8·04 kg   10. £15·45   11. 249·6 sec
12. 11·25 m   13. 7·36 m

Section 39
Division of decimals by numbers less than 10
Page 45

Exercise 1

1a. 8·4	b. 7·3	c. 10·05	d. 4·12
2a. 6·31	b. 0·63	c. 0·35	d. 2·11
3a. 4·5	b. 0·45	c. 0·045	d. 45·0
4a. 31·6	b. 14·1	c. 23·6	d. 2·36
5a. 5·04	b. 7·37	c. 62·4	d. 4·23
6a. 9·51	b. 51·4	c. 5·14	d. 12·61
7a. 11·5	b. 22·87	c. 123·13	d. 7·36
8a. 34·2	b. 23·8	c. 9·8	d. 12·31

9. £13·85   10. 4·78 cm   11. 0·56 kg   12. £5·25
13. 5·48 kg   14. 0·31 m   15. 21·45 sec

Section 40
Multiplication and division of decimals by 10, 100 and 1000
Page 46

Exercise 1

1. 34·1	2. 63·2	3. 73	4. 52·3	5. 58·4
6. 39·4	7. 383	8. 284	9. 483·6	10. 733·9
11. 10 846	12. 1000·4	13. 63·4	14. 900·4	15. 3451
16. 12 990	17. 3422	18. 28·47	19. 5483·7	20. 35 836
21. 5·72	22. 74·14	23. 8·398	24. 8·46	25. 0·648
26. 9·476	27. 0·239	28. 4·58	29. 0·836	30. 0·3561
31. 1·45003	32. 0·49002	33. 0·73725	34. 6·7946	35. 0·3746
36. 0·49007	37. 2·0737	38. 0·60047	39. 8·00005	40. 2·3592

Section 41
**Decimals
Revision
crossword**
Page 47

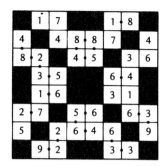

Section 42
**Multiplication
of fractions by
whole numbers**
Page 48

Exercise 1

1a. £3	**b.** 4 kg	**c.** 55 cm
2a. 3 kg	**b.** 4 l	**c.** 55 mm
3a. 3 cm	**b.** 5 ml	**c.** 6 g
4a. 3 m	**b.** 5 g	**c.** 7 l
5a. 8 kg	**b.** 10 mm	**c.** 20 ml
6a. 12 l	**b.** 18 t	**c.** 30 km
7a. $7\frac{1}{5}$ ml	**b.** 16 m	**c.** $23\frac{1}{5}$ g
8a. 12 mm	**b.** 21 kg	**c.** $31\frac{1}{5}$ l
9a. 12 cm	**b.** 18 l	**c.** $24\frac{9}{10}$ t
10a. 21 mm	**b.** 49 kg	**c.** 630 g
11a. 5 mm	**b.** 7 mm	**c.** 9 mm
12a. 15 l	**b.** 30 l	**c.** 40 l
13a. 2 mg	**b.** 3 mg	**c.** 5 mg
14a. 15 t	**b.** 20 t	**c.** 30 t
15a. 14 kg	**b.** 49 kg	**c.** 70 kg
16a. 6 t	**b.** 36 g	**c.** 35 mm

Exercise 2

1a. 6 t	**b.** 8 km	**c.** $13\frac{1}{2}$ l
2a. 2 m	**b.** 5 g	**c.** $6\frac{1}{4}$ ml
3a. 4 l	**b.** 6 cm	**c.** $10\frac{2}{3}$ kg
4a. 3 mm	**b.** 5 t	**c.** $8\frac{1}{5}$ ml
5a. 12 ml	**b.** 16 cm	**c.** 20 g
6a. 12 ml	**b.** 24 cm	**c.** $31\frac{1}{2}$ g
7a. 6 l	**b.** 10 kg	**c.** $11\frac{1}{5}$ mm
8a. 12 t	**b.** 21 cm	**c.** $16\frac{4}{5}$ l
9a. 6 mm	**b.** 12 kg	**c.** 18 ml
10a. $21\frac{3}{5}$ l	**b.** $32\frac{2}{5}$ m	**c.** $36\frac{9}{10}$ g
11a. 2 cm	**b.** 4 cm	**c.** 5 cm
12a. 15 g	**b.** 30 g	**c.** 40 g

13a. 2 m*l*   **b.** 7 m*l*   **c.** 3 m*l*

14a. 12 m   **b.** 24 m   **c.** 18 m

15a. 45 t   **b.** 25 t   **c.** 50 t

16a. 5 m   **b.** 27 kg   **c.** 55 cm

Section 43
Addition and subtraction of fractions
Page 50

Exercise 1

1a. 1   **b.** $\frac{1}{2}$ **c.** $\frac{3}{5}$ **d.** $\frac{3}{4}$

2a. $\frac{3}{4}$ **b.** 1   **c.** $\frac{9}{10}$ **d.** $\frac{4}{5}$

3a. 1   **b.** $\frac{2}{5}$ **c.** $\frac{2}{3}$ **d.** $\frac{5}{7}$

4a. $\frac{5}{6}$ **b.** 1   **c.** $1\frac{1}{3}$ **d.** $1\frac{1}{8}$

5a. $1\frac{1}{6}$ **b.** $1\frac{1}{10}$ **c.** $1\frac{1}{4}$ **d.** $1\frac{1}{50}$

Exercise 2

1a. $\frac{1}{2}$ **b.** $\frac{3}{5}$ **c.** $\frac{1}{5}$ **d.** $\frac{1}{5}$

2a. $\frac{2}{3}$ **b.** $\frac{1}{4}$ **c.** $\frac{1}{4}$ **d.** $\frac{1}{2}$

3a. $\frac{1}{4}$ **b.** $\frac{3}{4}$ **c.** $\frac{1}{5}$ **d.** $\frac{2}{5}$

4a. $\frac{3}{5}$ **b.** $\frac{3}{5}$ **c.** $\frac{4}{5}$ **d.** $\frac{31}{50}$

5a. $\frac{7}{25}$ **b.** $\frac{41}{100}$ **c.** $\frac{29}{100}$ **d.** $\frac{21}{50}$

Section 44
More addition and subtraction of fractions
Page 51

Exercise 1

1a. $\frac{3}{4}$ **b.** $1\frac{1}{4}$ **c.** $\frac{5}{6}$ **d.** $\frac{1}{2}$

2a. $\frac{5}{8}$ **b.** $\frac{7}{8}$ **c.** $\frac{2}{3}$ **d.** $\frac{5}{8}$

3a. $\frac{9}{10}$ **b.** $\frac{7}{8}$ **c.** $\frac{7}{8}$ **d.** $\frac{4}{5}$

4a. $1\frac{1}{10}$ **b.** $1\frac{1}{8}$ **c.** $1\frac{1}{2}$ **d.** $1\frac{3}{8}$

5a. $1\frac{2}{5}$ **b.** $1\frac{1}{16}$ **c.** $\frac{99}{100}$ **d.** $\frac{199}{1000}$

Exercise 2

1a. $\frac{1}{4}$ **b.** $\frac{1}{4}$ **c.** $\frac{1}{2}$ **d.** $\frac{1}{2}$

2a. $\frac{1}{8}$ **b.** $\frac{1}{8}$ **c.** $\frac{1}{10}$ **d.** $\frac{1}{3}$

3a. $\frac{1}{8}$ **b.** $\frac{1}{8}$ **c.** $\frac{1}{10}$ **d.** $\frac{1}{5}$

4a. $\frac{5}{8}$ **b.** $\frac{1}{8}$ **c.** $\frac{2}{5}$ **d.** $\frac{3}{8}$

5a. $\frac{1}{8}$ **b.** $\frac{3}{8}$ **c.** $\frac{3}{8}$ **d.** $\frac{5}{8}$

Section 45
Addition and subtraction of mixed numbers
Page 52

Exercise 1

1a. $6\frac{3}{5}$ **b.** $7\frac{3}{10}$ **c.** $5\frac{4}{5}$ **d.** $7\frac{2}{5}$

2a. 9   **b.** $10\frac{1}{2}$ **c.** 9   **d.** 16

3a. $10\frac{3}{5}$ **b.** $2\frac{5}{8}$ **c.** $3\frac{3}{4}$ **d.** $3\frac{3}{4}$

4a. $6\frac{3}{4}$ **b.** 4   **c.** 10   **d.** $8\frac{3}{10}$

5a. $4\frac{7}{10}$ b. $2\frac{1}{10}$ c. $2\frac{9}{10}$ d. $2\frac{1}{4}$

6a. $4\frac{1}{2}$ b. $7\frac{1}{2}$ c. $4\frac{5}{6}$ d. $4\frac{1}{6}$

7a. $3\frac{3}{8}$ b. $7\frac{1}{8}$ c. $6\frac{3}{8}$ d. $7\frac{5}{8}$

8a. $7\frac{1}{4}$ b. $3\frac{5}{12}$ c. $2\frac{11}{12}$ d. $3\frac{3}{4}$

Exercise 2

1a. $3\frac{1}{3}$ b. $2\frac{1}{2}$ c. $3\frac{1}{5}$ d. $3\frac{3}{5}$

2a. $3\frac{1}{2}$ b. $3\frac{1}{2}$ c. $4\frac{1}{4}$ d. $4\frac{2}{5}$

3a. $2\frac{1}{4}$ b. $4\frac{1}{8}$ c. $4\frac{1}{8}$ d. $5\frac{3}{10}$

4a. $3\frac{1}{5}$ b. $5\frac{1}{2}$ c. $5\frac{7}{10}$ d. $4\frac{3}{8}$

5a. $2\frac{1}{8}$ b. $2\frac{3}{8}$ c. $2\frac{3}{8}$ d. $4\frac{1}{2}$

Section 46
Percentages
Page 53

Exercise 1

1a. $\frac{1}{10}$ b. $\frac{1}{2}$ c. $\frac{3}{4}$ d. $\frac{1}{4}$

2a. $\frac{3}{10}$ b. $\frac{2}{5}$ c. $\frac{3}{5}$ d. $\frac{6}{25}$

3a. $\frac{1}{20}$ b. $\frac{9}{10}$ c. $\frac{4}{5}$ d. $\frac{9}{20}$

4a. $\frac{99}{100}$ b. $\frac{37}{100}$ c. $\frac{27}{100}$ d. $\frac{89}{100}$

5a. 1 b. 2 c. $1\frac{1}{2}$ d. 10 6. $\frac{55}{100} = \frac{11}{20}$

Exercise 2

1a. 9% b. 19% c. 37% d. 99%
2a. 30% b. 50% c. 70% d. 90%
3a. 40% b. 60% c. 80% d. 100%
4a. 25% b. 75% c. 15% d. 40%
5a. 65% b. 85% c. 12% d. 28%
6a. 56% b. 84% c. 6% d. 22%

Section 47
Using
percentages
Page 54

Exercise 1

1a. £5 b. £7 c. £1
2a. £3 b. £5 c. 5 km
3a. £2 b. £20 c. 10 women
4a. £15 b. £75 c. 30 men
5a. £27 b. £63 c. 1 workforce
6a. £24 b. £56 7. £6000 8. 250 pupils
9. £8 10. 21 pupils 11. £48 000 12. £640

Section 48
Fractions
Revision crossword
Page 55

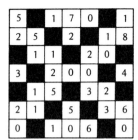

Using number
Revision test
Page 57

1a. nineteen   **b.** forty-two   **c.** two hundred and sixty-five
d. three thousand four hundred and one
e. thirteen thousand four hundred and nine

2a. 5   **b.** 12   **c.** 10   **d.** 10
   e. 8   **f.** 21   **g.** 16   **h.** 6

3a. 750   **b.** 36

4a. 96   **b.** 252   **c.** 832   **d.** 1494   **e.** 1705

5a. 15   **b.** 28   **c.** 4   **d.** 36   **e.** 3

6a. $\frac{3}{4}$ **b.** $\frac{1}{2}$ **c.** $\frac{4}{7}$ **d.** $\frac{3}{7}$ **e.** $\frac{1}{2}$

7a. $3\frac{1}{3}$ **b.** $4\frac{3}{4}$ **c.** $5\frac{2}{5}$ **d.** $6\frac{1}{5}$ **e.** $6\frac{2}{3}$

8a. 2·6, 4·5, 4·6, 5·4, 8·2   **b.** 6·7, 6·801, 6·866, 6·9
   c. 3·42, 3·421, 3·424, 3·43

9a. $\frac{2}{5}$ **b.** $\frac{13}{100}$ **c.** $\frac{21}{100}$ **d.** $\frac{13}{50}$ **e.** $\frac{8}{25}$
   f. $\frac{1}{2}$ **g.** $\frac{14}{125}$ **h.** $\frac{16}{125}$ **i.** $\frac{33}{100}$ **j.** $\frac{5}{8}$

10a. 0·2   **b.** 0·143   **c.** 0·4   **d.** 0·375   **e.** 0·7

11a. 2·6   **b.** 3·3   **c.** 5·12   **d.** 6·52   **e.** 8·38

12a. 3·3   **b.** 4·7   **c.** 1·4   **d.** 6·31   **e.** 2·13
   f. 1·60   **g.** 3·8   **h.** 1·7

13a. 1·5   **b.** 1·3   **c.** 6·41   **d.** 7·86   **e.** 10·7
   f. 12   **g.** 12   **h.** 3   **i.** 11   **j.** 9·81

14a. 11·4   **b.** 12·7   **c.** 5·28   **d.** 7·53   **e.** 8·950

15a. 4·3   **b.** 2·7   **c.** 2·42   **d.** 2·584   **e.** 2·474

16a. £18·40   **b.** 66·4 seconds   **c.** £16·42

17a. 64·2   **b.** 136   **c.** 141   **d.** 666·4   **e.** 1742·8
   f. 0·4864   **g.** 0·866   **h.** 3·842   **i.** 66·67   **j.** 42·866

18a. £7   **b.** £5   **c.** 7 kg   **d.** 26 m   **e.** 48 mm
   f. 400 *l*   **g.** £24   **h.** 28 g   **i.** 15 cm   **j.** 20 m

19a. $\frac{3}{5}$ **b.** $\frac{3}{4}$ **c.** $\frac{3}{5}$ **d.** $\frac{1}{2}$ **e.** $\frac{2}{5}$

20a. $7\frac{5}{8}$ b. $4\frac{3}{10}$ c. $8\frac{3}{10}$ d. $1\frac{3}{8}$ e. $3\frac{1}{2}$

21a. £6   b. £8·50   c. £4·50   d. £18   e. £24
  f. £6   g. £6   h. £132   i. £4   j. £10

22. £114 per week   23. £3000   24. 96

Part 2: Maths for Living

Section 49
Money
Page 61

Exercise 1

1a. £2·64   b. £4·14 .   5a. £400·65   b. £326·88
2a. £16·13   b. £42·20   6a. £5·04   b. £8·09
3a. £800·46   b. £6·99   7a. £22·01   b. £0·02
4a. £38·20   b. £10·35   8a. £16·08   b. £0·16

Exercise 2

1a. 0p   b. 1p   c. 3p   d. 9p
2a. 16p   b. 35p   c. 48p   d. 66p
3a. 75p   b. 86p   c. 90p   d. 96p
4a. 109p   b. 162p   c. 202p   d. 365p
5a. 469p   b. 578p   c. 682p   d. 916p
6a. 466p   b. 580p   c. 166p   d. 477p
7a. 584p   b. 294p   c. 191p   d. 36p
8a. 306p   b. 382p   c. 362p   d. 14p

Exercise 3

1a. £0·07   b. £0·15   c. £0·48   d. £0·96
2a. £1·25   b. £2·64   c. £3·81   d. £4·30
3a. £5·06   b. £6·45   c. £7·26   d. £8·36
4a. £9·45   b. £10·12   c. £11·41   d. £12·70
5a. £13·65   b. £14·14   c. £15·63   d. £16·58
6a. £18·32   b. £20·04   c. £21·04   d. £23·20
7a. £23·07   b. £23·02   c. £31·75   d. £32·40
8a. £44·16   b. £50·04   c. £63·86   d. £80·41

Exercise 4

1a. £2·39   b. £1·20   5a. £0·37   b. £0·50
2a. £12·91   b. £3·12   6a. £0·03   b. £0·01
3a. £7·02   b. £4·09   7a. £0·99   b. £10·06
4a. £1·01   b. £5·07   8a. £15·63   b. £20·50

Exercise 5

1a. seven pounds forty-five pence   b. five pounds forty-five pence
c. seven pounds thirty pence   d. two pounds seventy-one pence

2a. five pounds thirty-nine pence   **b.** three pounds ninety pence
  c. three pounds and five pence   **d.** five pounds and nine pence
3a. thirty-eight pence   **b.** sixty pence
  c. six pence   **d.** no pence
4a. three pounds twenty pence   **b.** eight pounds thirty-eight pence
  c. four pounds and six pence   **d.** two pounds forty-one pence
5a. four pounds eleven pence   **b.** six pounds twenty-two pence
  c. three pounds sixty-three pence   **d.** four pounds eighty-four pence
6a. two pounds twelve pence   **b.** eight pounds and four pence
  c. nine pounds seventeen pence   **d.** six pounds twenty-nine pence

7a. eleven pounds nineteen pence   **b.** thirteen pounds twenty-five pence
  c. fifteen pounds fifty-eight pence   **d.** eighteen pounds sixty-seven pence
8a. twenty-six pounds sixty-nine pence **b.** thirty-one pounds seventy-five pence
  c. forty-two pounds eighty pence   **d.** fifty pounds ninety-nine pence

Section 50
**Addition and
subtraction of
money**
Page 63

Exercise 1

1a. £2·97	**b.** £7·94	**c.** £18·99
2a. £3·93	**b.** £9·91	**c.** £29·90
3a. £5·18	**b.** £5·59	**c.** £1·09
4a. £9·35	**b.** £5·30	**c.** £21·34
5a. £3·79	**b.** £9·91	**c.** £18·62

Exercise 2

1a. £3·11	**b.** £3·20	**c.** £11·22
2a. £4·14	**b.** £0·06	**c.** £23·86
3a. £5·76	**b.** £6·09	**c.** £18·89
4a. £2·08	**b.** £5·19	**c.** £2·67
5a. £6·67	**b.** £0·99	**c.** £5·80
6a. £3·68	**b.** £9·45	**c.** £1·79

Section 51
**Multiplication
and division
of money**
Page 64

Exercise 1

1a. £6·82	**b.** £9·69	**c.** £8·10	**d.** £6·25
2a. £13·83	**b.** £6·96	**c.** £11·12	**d.** £5·11
3a. £36·36	**b.** £23·56	**c.** £29·28	**d.** £36·72
4a. £2·88	**5.** £13·59		

Exercise 2

1a. £2·13	**b.** £41·13	**c.** £1·02
2a. £0·60	**b.** £4·00	**c.** £2·13
3a. £3·15	**b.** £3·10	**c.** £0·90
4a. £0·63	**b.** £1·02	**c.** £2·86

5. £0·62   **6.** 0·45 kg   **7.** £5·63   **8.** 2·15 m   **9.** 152·4 kg

Section 52
**Pay and
overtime**
Page 65

Exercise 1

time and a quarter	time and a half	double time
225	270	360
250	300	400
275	330	440
300	360	480
325	390	520
350	420	560
375	450	600
400	480	640
425	510	680
450	540	720
475	570	760
500	600	800
525	630	840
550	660	880
575	690	920
600	720	960
625	750	1000
650	780	1040
675	810	1080
700	840	1120
725	870	1160
750	900	1200

Section 53
Gross wages
Page 66

Exercise 1

1. £100·00 2. £144·00 3. £168·00 4. £228·00 5. £79·50
6. £112·50 7. £119·00 8. £171·00 9. £85·40 10. £135·28

Exercise 2

1. £110·00 2. £168·00 3. £184·00 4. £216·00 5. £76·00
6. £74·00 7. £125·60 8. £106·40 9. £77·70 10. £82·60
11. £103·68 12. £117·60 13. £148·32 14. £192·44

Section 54
**Gross wages:
overtime**
Page 67

Exercise 1

1. £94·50 2. £120·00 3. £152·00 4. £189·00 5. £252·00
6. £154·00 7. £241·50 8. £198·00 9. £151·25 10. £288·00

Section 55
**Gross wages:
pay slip**
Page 67

Exercise 1

1.

MON	11	8	3	1600p	750p	£23·50
TUES	8	8	0	1600p		16·00
WED	10	8	2	1600p	500p	21·00
THURS	12	8	4	1600p	1000p	26·00
FRI	8	8	0	1600p		16·00
SAT	4	0	4		1200p	12·00
					gross wage	£114·50

2.

MON	10	8	2	1920p	600p	£25·20
TUES	8	8	0	1920p		19·20
WED	11	8	3	1920p	900p	28·20
THURS	10	8	2	1920p	600p	25·20
FRI	10	8	2	1920p	600p	25·20
SAT						
SUN	6	0	6		2880p	28·80
					gross wage	£151·80

3.						
MON	10	8	2	2880p	900p	£37·80
TUES	12	8	4	2880p	1800p	46·80
WED	14	8	6	2880p	2700p	55·80
THURS	12	8	4	2880p	1800p	46·80
FRI						
SAT	10		10		5400p	54·00
					gross wage	£241·20

Section 56
Piecework
Page 69

Exercise 1

1a. £1·20 b. £3·00 6a. £23·28 b. £32·34
2a. £6·00 b. £12·50 7a. £0·10 b. £0·30
3a. £20·40 b. £28·00 8a. £0·60 b. £1·00
4a. £1·50 b. £4·32 9a. £1·70 b. £2·16
5a. £9·00 b. £15·40 10a. £2·64 b. £3·36

Exercise 2

Armstrong	£11·52	Khan	£18·24	Singh	£9·00
Clark	£11·10	Leniham	£17·55	Sheth	£9·00
Dawson	£8·96	Miller	£11·16	Winter	£14·00
Elliot	£17·50	Parekh	£21·06	Yardley	£9·60
Gough	£17·08	Rose	£15·60	Young	£16·00

Section 57
Wages:
salaries
Page 70

Exercise 1

1. £500 2. £1100 3. £855 4. £1230 5. £625
6. £1245 7. £688 8. £1425 9. £655 10. £1260

Exercise 2

1. £18 660 2. £17 760 3. £10 140 4. £13 080 5. £12 540
6. £20 160 7. £11 760 8. £10 080 9. £15 360 10. £16 200

Section 58
Net wage
Page 71

Exercise 1

1. £93.99 2. £115.69 3. £93.57 4. £87.82 5. £104.69
6. £88·98 7. £110·19 8. £70·92 9. £99·20 10. £121·55
11. £114·55 12. £96·35 13. £109·11 14. £106·26 15. £109·10

Section 59
Income tax:
P.A.Y.E.
Page 72

Exercise 1

1a. £2205 b. £2335 c. £700·50
2a. £2205 b. £7555 c. £2266·50
3a. £3455 b. £2775 c. £832·50
4a. £3555 b. £4945 c. £1483·50
5a. £3555 b. £5945 c. £1783·50
6a. £2690 b. £6010 c. £1803·00
7a. £2405 b. £8925 c. £2677·50

Section 60
Unemployment and supplementary benefit
Page 73

Exercise 1

1. £45·55   **2.** £14·35   **3.** 18   **4.** £10·05   **5.** £27·55

Exercise 2

1. £95·75   **2.** £47·25   **3.** £59·40   **4.** £40·95   **5.** £114·60
6. £109·65   **7.** £27·65   **8.** £83·60   **9.** £75·70   **10.** £100·60

Section 61
Banking: paying in
Page 75

Exercise 1

1. £53·10   **2.** £65·43   **3.** £119·63   **4.** £76·43
5. £133·02   **6.** £150·96   **7.** £258·23   **8.** £189·95

Section 62
Banking: cheques
Page 76

Exercise 1

1. six pounds 86   £6·86
2. twenty-three pounds 60   £23·60
3. eighty pounds only   £80·00
4. thirty-six pounds 58   £36·58
5. five hundred and sixty pounds 55   £560·55
6. sixty-eight pounds 60   £68·60
7. forty-nine pounds 55   £49·55
8. fifty pounds only   £50·00
9. nineteen pounds 50   £19·50
10. eight thousand seven hundred and fifty pounds only   £8750·00
11. thirty-two pounds 65   £32·65
12. thirty-five pounds only   £35·00

Section 63
Banking: statements
Page 77

Exercise 1

1. £647·42   **2.** £647·42   **3.** £560·42   **4.** £377·54   **5.** £750·04
6. £482·16   **7.** monthly wage   **8.** money paid in   **9.** £718·04
10. £718·04   **11.** £900   **12.** £913·80   **13.** £144·00

Section 64
**Banking:
using a
statement**
Page 78

Exercise 1

1.

PAYMENTS	RECEIPTS	DATE	BALANCE
		16 Aug	£644·84
10·16		18	634·68
12·42		21	622·26
25·00		25	597·26
74·53		28	522·73
	450·00	30	972·73
25·00		3 Sept	947·73
50·00		6	897·73
124·16		8	773·57
12·66		11	760·91
84·50		18	676·41
10·16		18	666·25
12·42		21	653·83
74·53		28	579·30
	450·00	30	1029·30
80·00		4 Oct	949·30
	500·00	6	1449·30

2a. £634·68   **b.** £972·73   **c.** £1449·30   **3.** Car instalment
4. £894·36   **5.** £5400   **6.** 641 290   **7.** Gas, rates, car
8. £122·11   **9a.** £393·43   **b.** No — probably more

Section 65
**Banking:
credit cards**
Page 79

Exercise 1

1a. a meal   **b.** £1·25   **c.** £39·43   **d.** £15·00
e. 30th Jan., 1982   **f.** £7·00

Section 66
**Post Office:
letters and
parcels**
Page 80

Exercise 1

1. 51   **2.** 60   **3.** 144   **4.** 88
5. 266p   **6.** 368p   **7.** £11·28   **8.** £6·69
9. £13·25   **10.** £20·65   **11.** £25·20   **12.** £22·40
13a. 17p   **b.** 18p   **c.** 69p   **d.** £1·82   **e.** £2·44
14. 24p   **15.** £1·70   **16.** £15·51   **17.** £2·70   **18.** 8; 15p change
19. 8; 4p change   **20.** £4·34

Section 67
**Post Office:
postal orders**
Page 81

Exercise 1

1. 60p   **2.** 89p   **3.** £1·31   **4.** £1·83   **5.** £6.16
6. £2·63   **7.** £3·55   **8.** £9·75   **9.** £15·50   **10.** £19·43

Section 68
Telephones
Page 82

Exercise 1

1a. 6p   **b.** 6p   **c.** 6p
2a. 46p   **b.** 58p   **c.** 17p

3a. 63p **b.** 46p **c.** 23p **4.** 40p
5. 17p **6.** 58p **7.** About three times cheaper

Section 69
V.A.T.
Page 83

Exercise 1

1. £23 **2.** £34·50 **3.** £115 **4.** £207 **5.** £287·50
6. £345 **7.** £517·50 **8.** £690 **9.** £1380 **10.** £28·75

Section 70
Calculating
V.A.T.
Page 84

Exercise 1

1. 36p, £2·76 **2.** £2·75, £21·05 **3.** £11·22, £86·02
4. £1·87, £14·36 **5.** £3·39, £25·95 **6.** £7·50, £57·49
7. £9·87, £75·62 **8.** £13·50, £103·49 **9.** £6·94, £53·18
10. £14·99, £114·89 **11.** £3·70, £28·36 **12.** £13·11, £100·51
13. £1·35, £10·34 **14.** £2·38, £18·23 **15.** £1·89, £14·51

Section 71
Car insurance:
premiums
Page 85

Exercise 1

1. £436 **2.** £664 **3.** £707 **4.** £795 **5.** £944
6. £950 **7.** £627 **8.** £716 **9.** £1568 **10.** £2684

Section 72
Car insurance:
no-claims bonus
Page 86

Exercise 1

1. £306 **2.** £970 **3.** £594 **4.** £357·50 **5.** £1460
6. £216·80 **7.** £708 **8.** £940·80 **9.** £348·75 **10.** £497

Section 73
Hire purchase
Page 87

Exercise 1

1a. £8·50 **b.** £1·50 **6a.** £422 **b.** £72
2a. £60 **b.** £15 **7a.** £3208 **b.** £708
3a. £23·10 **b.** £5·10 **8a.** £4586 **b.** £1386
4a. £91 **b.** £16 **9a.** £213·80 **b.** £43·80
5a. £180 **b.** £25 **10a.** £952·40 **b.** £102·40

Section 74
Rent
Page 88

Exercise 1

1. £19·20 **2.** £2·40 **3.** £21·60 **4.** £86·40 **5.** £43·20
6. £960·00 **7.** £120 **8.** £1080 **9.** A. Brown **10.** She was out
11. No
12.

1 Dec	16·80	2·64	19·44	19·44
8	16·80	2·64	19·44	
15	16·80	2·64	19·44	38·88	38·88
22	16·80	2·64	19·44	
29	16·80	2·64	19·44	38·88	38·88

13. £84·00 **14.** £13·20 **15.** £132·00 **16.** £840

Section 75
House purchase
Page 89

Exercise 1

1a. £88·00   **b.** £1056   **2.** £216   **3a.** £21 000   **b.** £225
4a. £6000   **b.** £225   **5a.** £10 000   **b.** £375
6a. £432   **b.** £5184   **7a.** £11 000   **b.** £550   **c.** £6600
8a. £625   **b.** £7500   **c.** £167 000 (including the deposit)
9a. £15 000   **b.** £648   **c.** £7776   **d.** £209 400
10a. £45 000   **b.** £562·50   **c.** £6750   **d.** £180 000

Section 76
**Household
insurance**
Page 91

Exercise 1

1a. D   **b.** C   **c.** B   **d.** E   **e.** F
2a. £21·50   **b.** £19·25   **c.** £16·25   **d.** £23·00   **e.** £26·75
3. £17·30   **4.** £13·50   **5.** £29·45   **6.** £17·50   **7.** £14·75
8. £13·50   **9.** £27·35   **10.** £35·30   **11.** £19·25   **12.** £18·75
13. £25·25   **14.** £21·75   **15.** £16·25   **16.** £17·30

Section 77
Life assurance
Page 92

Exercise 1

1. £268·50   **2.** £106·50   **3a.** £366   **b.** £30·50
4a. £109   **b.** £9·08   **5a.** £226   **b.** £18·83
6a. £380·40   **b.** £31·70   **7a.** £618·40   **b.** £51·53
8a. £950   **b.** £79·17

Section 78
Electricity
Page 93

Exercise 1

1. 9 kWh   **2.** 16 kWh   **3.** 15 kWh   **4.** 3 kWh
5. 2 kWh   **6.** 2 kWh   **7.** 3 kWh   **8.** 1 kWh

Exercise 2

1. 96p   **2.** £1·80   **3.** 24p   **4.** 36p   **5.** £1·08
6. £1·44   **7.** 72p   **8.** 43p   **9.** 24p   **10.** £1·08

Section 79
**Reading the
meter**
Page 94

Exercise 1

1. 6530 units   **2.** 9606 units   **3.** 3573 units
4. 5237 units   **5.** 7633 units

Section 80
**Electricity
bills**
Page 95

Exercise 1

1. 00610, 01003, 01218, 00753, 00609
2. 01320   **3.** 02105   **4.** 01320   **5.** 00785
6. More electricity is used in Winter   **7.** 9th January

Exercise 2

1. £27·00   **2.** £72·00   **3.** £37·44   **4.** £165·06   **5.** £43·74
6. £45·30   **7.** £38·52   **8.** £65·52   **9.** £57·00   **10.** £86·58

Section 81
Gas bills
Page 96

Exercise 1

1. £92·80 **2.** £126·58 **3.** £454·00 **4.** £212·80 **5.** £272·80
6. £904·00 **7.** £145·58 **8.** £126·80 **9.** £198·00 **10.** £697·90

Section 82
Foreign currency
Page 97

Exercise 1

1. 760 francs **2.** 4800 drachmas **3.** 52 dollars **4.** 110 000 lire
5. 21 000 pesetas **6.** 456 marks **7.** 465 sw fr **8.** 4200 pesetas
9. 325 dollars **10.** 226 600 lire

Section 83
Holidays abroad
Page 98

Exercise 1

1. £12·00 **2.** £10·00 **3.** £58·00 **4.** £24·00
5. £26·00 **6.** £78·00 **7.** £196·00 **8.** £78·00
9. £208·00 **10.** £63·00 **11.** £176·00 **12.** £146·00

Section 84
Booking a holiday
Page 99

Exercise 1

1. £773 **2.** £1459·50 **3.** £493·00 **4.** £1552·50
5. £853·00 **6.** £1169·50 **7.** £1546 **8.** £1706
9. £1026 **10.** £1502·50 **11.** £1466 **12.** £1559

Section 85
Time: the calendar
Page 100

Exercise 1

1. **2.** Tuesday **3.** Monday **4.** Yes **5.** 25th Dec.
6. 5 **7.** 4th Jan. **8.** 8 **9.** 16 days **10.** Sunday
11. 23 days **12.** 4th Sept.

Section 86
Time: the clock
Page 101

Exercise 1

1a. ten past twelve (noon) **b.** fourteen minutes past seven
c. twenty minutes past eight
d. twenty-five minutes past one
2a. half past two **b.** half past ten
c. half past nine **d.** half past four
3a. quarter past eleven **b.** quarter past five
c. quarter to eleven **c.** quarter to seven
4a. ten minutes to four **b.** one minute to three
c. twenty-five minutes to eleven **d.** twenty minutes to five
5a. twenty-nine minutes to nine **b.** fourteen minutes to two
c. five minutes to six **d.** twenty-four minutes to eight

Exercise 2

1a. 6·20 **b.** 2·16 **9a.** 6·56 **b.** 2·31
2a. 1·23 **b.** 12·25 **10a.** 12·45 **b.** 12·59
3a. 11·30 **b.** 3·30 **11a.** 3·44 **b.** 5·11
4a. 8·15 **b.** 5·15 **12a.** 9·40 **b.** 5·20
5a. 8·45 **b.** 9·45 **13a.** 11·35 **b.** 1·29
6a. 6·50 **b.** 5·40 **14a.** 1·31 **b.** 8·06

7a. 3·35   **b.** 11·55   **15a.** 6·54   **b.** 4·49
8a. 7·42   **b.** 1·48   **16a.** 10·01   **b.** 9·59

Section 87
**Time: a.m.
and p.m.**
Page 102

Exercise 1

1a. 9.00 a.m.   **b.** 10.20 a.m.   **2a.** 1.22 p.m.   **b.** 3.45 p.m.
3a. 5.40 a.m.   **b.** 8.10 p.m.   **4a.** 10.20 p.m. **b.** 3.20 a.m.
5a. 7.30 a.m.   **b.** 6.05 a.m.   **6a.** 12.20 p.m. **b.** 12.50 a.m.

Section 88
**Time: the
24-hour clock**
Page 103

Exercise 1

1a. 10.20   **b.** 11.15   **c.** 22.20
2a. 16.15   **b.** 16.30   **c.** 23.47
3a. 03.15   **b.** 07.55   **c.** 09.15
4a. 21.50   **b.** 20.08   **c.** 01.00
5a. 11.00   **b.** 05.30   **c.** 13.00
6a. 00.30   **b.** 18.45   **c.** 10.50
7a. 19.25   **b.** 19.10   **c.** 18.40
8a. 08.45   **b.** 12.45   **c.** 21.34
9a. 21.48   **b.** 23.16   **c.** 11.45
10a. 10.55   **b.** 06.59   **c.** 00.50

Exercise 2

1a. 11.58 a.m.   **b.** 2.55 a.m.   **c.** 4.22 a.m.
2a. 4.50 a.m.   **b.** 5.34 a.m.   **c.** 5.50 p.m.
3a. 3.45 a.m.   **b.** 10.50 p.m.   **c.** 4.48 p.m.
4a. 2.40 p.m.   **b.** 11.25 p.m   **c.** 3.35 a.m.
5a. 6.55 p.m.   **b.** 8.45 a.m.   **c.** 12.35 p.m.
6a. 9.40 a.m.   **b.** 1.30 a.m.   **c.** 12.24 a.m.
7a. 2.50 a.m.   **b.** 1.35 p.m.   **c.** 12.50 a.m.
8a. 3.30 p.m.   **b.** 3.40 p.m.   **c.** 3.55 a.m.
9a. 7.20 p.m.   **b.** 4.25 a.m.   **c.** 2.52 p.m.
10a. 9.45 p.m.   **b.** 5.37 a.m.   **c.** 1.38 a.m.

Section 89
**Time:
subtraction**
Page 104

Exercise 1

1a. 5 hrs 10 mins   **b.** 5 hrs 15 mins
    c. 4 hrs 15 mins   **d.** 4 hrs 25 mins
2a. 3 hrs 40 mins   **b.** 4 hrs 55 mins
    c. 5 hrs 37 mins   **d.** 5 hrs 18 mins
3a. 1 hr 15 mins   **b.** 3 hrs 20 mins
4a. 3 hrs 05 mins   **b.** 5 hrs 15 mins
5a. 2 hrs 30 mins   **b.** 1 hr 55 mins
6a. 2 hrs 40 mins   **b.** 5 hrs 15 mins
7. 10 hr 50 min   **8.** 10 hrs 31 mins
9. 13 hrs 55 mins   **10.** 14 hrs 20 mins

Section 90
**Time: using
the 24-hour
clock**
Page 105

Exercise 1

1. R   2. R   3. R   1 hr 36 mins   4. 48 mins
5. Winchester & Basingstoke   6. T   7. 30 mins
8. 2 hrs 10 mins   9. 2 hrs 40 mins   10. 1 hr 22 mins
11. 1 hr 04 mins   12. 56 mins   13. 2 hrs 18 mins
14. 1 hr 54 mins   15. 2 hrs 38 mins   16. 1 hr 26 mins
17. 3 hrs 08 mins   18. 52 mins   19. 1 hr 44 mins
20. 2 hrs 18 mins

Section 91
**Using
timetables**
Page 106

Exercise 1

1. 1 hr 50 mins   2. 4 hrs 32 mins   3. 4 hrs 34 mins
4. 1 hr 58 mins   5. 1 hr 56 mins   6. 1 hr 32 mins
7. 6 hrs 02 mins   8. 10 hrs 20 mins   9. 10 hrs 16 mins
10. 4 hrs 28 mins   11. 231 miles   12. 206 miles
13. 121 miles 14. 506 miles 15. 43 miles 16. 522 miles 17. 158 miles
18. 257 miles 19. 157 miles 20. 392 miles

Section 92
**Distance,
speed and time**
Page 107

Exercise 1

1. 5 hrs   2. 5 hrs   3. 10 hrs   4. 5 hrs   5. 9 hrs
6. 4 hrs   7. 10 hrs   8. 10 hrs   9. 8 hrs   10. 4 hrs

Exercise 2

1. 50 hrs   2. 30 mins   3. 5 hrs   4. 5 hrs   5. 50 hrs

Section 93
**Speed, distance
and time**
Page 108

Exercise 1

1. 100 km   2. 180 km   3. 360 km   4. 550 km   5. 840 km
6. 984 km   7. 10 km/hr   8. 20 km/hr   9. 20 km/hr   10. 30 km/hr
11. 30 km/hr 12. 32 km/hr

Exercise 2

1. 20 km/hr   2. 15 km/hr   3. 100 km/hr 4. 96 km   5. 120 metres
6. $2\frac{1}{2}$ hrs   7. $5\frac{1}{2}$ hrs

Section 94
Simple interest
Page 109

Exercise 1

1. £9   2. £16   3. £30   4. £120   5. £280
6. £288   7. £500   8. £675   9. £2000   10. £4500

Exercise 2

1. £19·60   2. £63·00   3. £130·00   4. £127·00   5. £55·00
6. £136·50   7. £156·66   8. £155·20   9. £339·75   10. £945·60

Section 95
Compound interest
Page 110

Exercise 1

1. £102·50 2. £247·20 3. £434·70 4. £832·00 5. £571·20
6. £1504·80 7. £548·10 8. £2100 9. £3816 10. £6450

Section 96
Profit and loss
Page 111

Exercise 1

1. 20% profit 2. 50% profit 3. 75% loss 4. 50% loss
5. 25% profit 6. 25% profit 7. 100% profit 8. 100% loss
9. 20% loss 10. 75% loss

Section 98
Petty cash accounts
Page 113

Exercise 1

June 17	*cash received*	£13·13
June 24	*cash received*	£12·89
June 28	*balance*	£10·70

Section 99
Running balances
Page 114

Exercise 1

1. £645·47 £677·91 £553·49 £537·89 £529·39
   £555·99 £537·99 £553·64 £547·16
2. £553·49 3. No

Exercise 2

Final balance £251·22

Section 100
Cash accounts
Page 115

	Exercise 1	Exercise 2	Exercise 3	Exercise 4
Total income	£1191·09	£276·29	£1051·32	£913·39
Total expenditure	£544·07	£215·48	£220·67	£344·00
Balance	£647·02	£60·81	£830·65	£569·39

Section 101
Maths for living revision crossword
Page 117

7	5	■	7	6	■	1	2
5	■	8	5	2	0	■	4
■	2	■	6	0	■	1	■
1	1	6	■	■	7	4	4
1	5	0	■	■	2	7	0
■	0	■	1	4	■	0	■
5	■	1	8	1	5	■	9
4	9	■	2	6	■	2	0

Maths for living: Revision test
Page 119

1a. 220p b. 331p c. 1250p d. 207p
e. 10p f. 6p g. 300p h. 2000p

2a. £1·22 b. £3·06 c. £2·80 d. £0·28
e. £28·00 f. £20·80 g. £0·02 h. £0·10

3a. £3·58   **b.** £4·01   **c.** £21·62   **d.** £20·48

4a. £0·11   **b.** £1·47   **c.** £4·05   **d.** £12·88

5a. £15·39   **b.** £11·15   **c.** £18·24   **d.** £27·54

6a. £1·18   **b.** £1·57   **c.** £4·05   **d.** £1·35

7. 12p   **8.** 48p   **9.** £118.00
10. £21·00   **11.** £625   **12.** £40·25
13. £70   **14.** £2880

15a. 10.00 a.m.   **b.** 10.00 p.m.   **c.** 8.20 a.m.
    d. 3.00 a.m.   **e.** 11.30 p.m.   **f.** 12.30 p.m.

16a. 19.40   **b.** 07.40   **c.** 23.20   **d.** 11.00
    e. 12.17   **f.** 00.17   **g.** 08.12   **h.** 02.00

17a. 120 km   **b.** 3 hr   **c.** 40 km/hr

18a. 2 hr 45 m   **b.** 3 hr 27 m   **c.** 6 hr 26 m   **d.** 10 hr 25 m

19. 2 hr 36 min   **20.** Train X, by 6 min

21a. 7 hrs   **b.** 85 km per hour   **c.** 480 km

22a. £410   **b.** £1260   **c.** £1935

23a. 20% loss   **b.** 25% loss   **c.** 400% profit

Part 3: Geometry

Section 102
Measuring
Page 121

Exercise 1

1. 3 cm 6 mm   **2.** 10 cm 5 mm   **3.** 7 cm 7 m
4. 5 cm 1 mm   **5.** 11 cm 2 mm   **6.** 4 cm 4 m

Section 103
Estimating
Page 122

Exercise 1

1. 5·8 cm   **2.** 8·2 cm   **3.** 11·0 cm   **4.** 3·5 cm   **5.** 7·0 cm
6. 12·3 cm   **7.** 5·6 cm   **8.** 7·2 cm   **9.** 6·5 cm   **10.** 7·4 cm

Section 105
Angles
Page 124

Exercise 1

1a. 180°   **b.** 90°   **2a.** 90°   **b.** 360°   **3.** 360°
4. 270°   **5.** 720°   **6.** 180°   **7.** 90°   **8.** 360°
9. 0°   **10.** 270°   **11.** 270°   **12.** 180°

Section 106
Angle measure
Page 125

Exercise 1

1. 57°   **2.** 30°   **3.** 76°   **4.** 113°   **5.** 90°
6. 103°   **7.** 90°   **8.** 65°   **9.** 130°   **10.** 54°

Section 107
**Angle estimation
and measurement**
Page 127

Exercise 1

1. 43°   2. 66°   3. 30°   4. 105°   5. 33°
6. 110°   7. 88°   8. 55°   9. 140°   10. 67°

Section 108
**Angles: acute,
obtuse and
reflex**
Page 129

Exercise 1

1. Acute angles are 1, 2, 3, 8, 10
2. Obtuse angles are 4, 6, 9.

Exercise 2

1. 328°   2. 230°   3. 334°   4. 192°   5. 310°   6. 355°

Section 111
**Triangle
measurement**
Page 133

Exercise 1

1. A = 80°   B = 40°   C = 60°   2. X = 60°   Y = 50°   Z = 70°
3. L = 70°   M = 80°   N = 30°   4. B = 40°   C = 50°   D = 90°
5. J = 80°   K = 51°   L = 49°   6. P = 60°   Q = 60°   R = 60°
7. R = 70°   S = 80°   T = 30°   8. X = 20°   Y = 120° Z = 40°
9. A = 70°   B = 40°   C = 70°   10. D = 30°   E = 30°   F = 120°

Section 112
**Angle sum of
a triangle**
Page 135

Exercise 1

1. 80°   2. 40°   3. 50°   4. 40°   5. 40°
6. 60°   7. 135°   8. 60°   9. 45°

1. 59°   2. 55°   3. 62°   4. 29°   5. 26°
6. 95°   7. 101°   8. 100°   9. 40°   10. 52°
11. 44°   12. 40°

Section 113
**Special
triangles**
Page 137

Exercise 1

1. A, C, E, F
2. A = 73° 73° 34°
   C = 45° 45° 90°
   E = 61° 61° 58°
   F = 83° 83° 16°
   Two angles are the same.

Exercise 2

1. B, C, E
2. B = 60° 60° 60°
   C = 60° 60° 60°
   E = 60° 60° 60°
   Each angle is always 60°

Section 114
Squares and rectangles
Page 138

Exercise 1

2c. 4 cm 2 mm	d. Yes	3c. 5 cm	d. Yes
4c. 7 cm 1 mm	d. Yes	e. 90°	f. No
5c. 67 mm d. Yes	e. 53°	f. 127°	g. Yes

Section 115
Squares and rectangles: perimeter
Page 139

Exercise 1

1. 11·8 cm   2. 14·0 cm   3. 13·8 cm   4. 10 cm   5. 11·4 cm
6. 12·0 cm   7. 180 mm   8. 208 mm   9. 4 cm   10. 30 m

Section 116
Squares and rectangles: area
Page 140

Exercise 1

1. 15 cm^2 2. $10·5 \text{ cm}^2$ 3. 8 cm^2 4. 14 cm^2 5. $6·25 \text{ cm}^2$
6. 6 cm^2 7. 81 m^2 8. 100 tiles

Exercise 2

1. 13 cm^2 2. 12 cm^2 3. 8 cm^2 4. 8 cm^2 5. 16 cm^2
6. $10·2 \text{ cm}^2$ 7. 12 cm^2 8. 8 cm^2 9. 8 cm^2
10. $100 \text{ mm}^2 = 1 \text{ cm}^2$

Section 117
Triangles: area
Page 142

Exercise 1

1. b or d
2a. 1 cm^2
 b. 2 cm^2
 c. $1\frac{1}{2} \text{ cm}^2$
 d. 2 cm^2
 e. 1 cm^2
3. The areas are the same because the triangles have the same base and height

Exercise 2

1. Q is (3 −1)   R is (−2 −1)
2. —
3. 15
4. 7½
5. Triangle PQR has half the area of rectangle PQRS

Section 118
Triangles: area II
Page 143

Exercise 1

1. $3·9 \text{ cm}^2$ 2. 2 cm^2 3. $3·3 \text{ cm}^2$ 4. 3 cm^2 5. $3·45 \text{ cm}^2$
6. $2·7 \text{ cm}^2$ 7. $2·625 \text{ cm}^2$ 8. 4 cm^2 9. $3·2 \text{ cm}^2$

Exercise 2

1. 2 cm^2	**2.** 15 cm^2	**3.** 48 cm^2	**4.** 66 cm^2	**5.** 3·6 cm^2
6. 10 cm^2	**7.** 9·2 cm^2	**8.** 8·7 cm^2	**9.** 9·9 cm^2	**10.** 16·45 cm^2
11. 11·2 cm^2	**12.** 13·2 cm^2			

Section 119
**Area and
perimeter**
Page 145

Exercise 1

1. 18 cm, 16 cm^2	**2.** 12 cm, 6·5 cm^2	**3.** 13 cm, 7·5 cm^2
4. 20 m, 16 m^2	**5.** 16 m, 16 m^2	**6.** 18 m, 16 m^2

Exercise 2

1. 17·5 cm^2	**2.** 11 cm^2	**3.** 6 cm^2
4. 12 cm^2	**5.** 12 cm^2	**6.** 7 cm^2

Section 120
Circles
Page 146

Exercise 1

1. 2 cm	**2.** 1 cm	**3.** 1·5 cm
4. 1·35 cm	**5.** 0·9 cm	**6.** 1·8 cm

7. DIAMETERS: (1) 4 cm (2) 2 cm (3) 3 cm (4) 2·7 cm
(5) 1·8 cm (6) 3·6 cm

Section 121
**Circles:
circumference**
Page 147

Exercise 1

1a. 6·28 mm **b.** 15·7 mm   **2a.** 31·4 mm **b.** 18·84 cm
3a. 28·26 cm **b.** 37·68 cm   **4a.** 12·56 mm **b.** 25·12 mm
5a. 43·96 cm **b.** 50·24 cm
6a. 91·06 mm **b.** 81·64 mm **c.** 62·8 mm   **d.** 72·22 mm
7. 31·4 m   **8.** 1·57 m   **9.** 31·4 cm

Section 122
**Circles:
area**
Page 148

Exercise 1 (Answers given to 3 s.f.)

1a. 3·14 mm^2 **b.** 50·2 cm^2 **2a.** 78·5 cm^2 **b.** 113 mm^2
3a. 200 mm^2 **b.** 254 cm^2 **4a.** 12·6 m^2 **b.** 28·3 cm^2
5a. 154 cm^2 **b.** 314 mm^2
6a. 660 mm^2 **b.** 530 mm^2 **c.** 314 mm^2 **d.** 415 mm^2
7. 706 m^2 **8.** 2460 cm^2 **9.** 11300 cm^2 **10.** 4070 cm^2
11. 3·14 cm^2 **12.** 28·2 cm^2 **13.** 283 cm^2 **14.** 201 cm^2

Section 123
Co-ordinates
Page 149

Exercise 1

1. (4, −4)	**6.** (8, −9)	
2. (−5, 6)	**7.** (7, −1)	
3. (−6, −8)	**8.** (−7, −5)	
4. (−5, −5)	**9.** (5, −2)	
5. (8, 3)	**10.** (−5, −4)	

Exercise 2

1.	Lion's Den	**6.**	Waterfall
2.	Lion's Den	**7.**	Well
3.	Tin Mine	**8.**	Lighthouse
4.	Black Jack's Cabin	**9.**	Pond
5.	Potholes	**10.**	Salt Lake

Exercise 3

1. 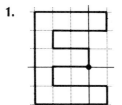 **2.** 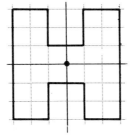 **3.**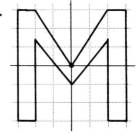

4. 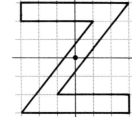 **5.** 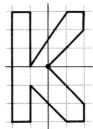 **6.**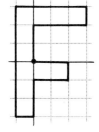

Section 124
**Maps: scale
and distance**
Page 151

Exercise 1

1. 230 km	**2.** 140 km	**3.** 190 km	**4.** 290 km	**5.** 410 km
6. 130 km	**7.** 320 km	**8.** 760 km	**9.** 400 km	**10.** 40 km

Section 125
**Maps:
direction**
Page 152

Exercise 1

1. Brighton	**2.** Nottingham	**3.** Southampton F.C.
4. Manchester	**5.** Nottingham	**6.** Preston
7. Brighton	**8.** Liverpool	**9.** Oxford
10. Cardiff City	**11.** Southampton F.C.	**12.** Liverpool
13. Glasgow	**14.** 450 km	**15.** 220 km

Section 126
Bearings
Page 153

Exercise 1

1. 030°	**2.** 090°	**3.** 130°	**4.** 150°	**5.** 110°	**6.** 180°

Exercise 2

1. 270°   **2.** 310°   **3.** 255°   **4.** 230°   **5.** 200°   **6.** 180°

Exercise 3

1. 060°   **2.** 045°   **3.** 045°   **4.** 120°   **5.** 120°
6. 275°   **7.** 240°   **8.** 328°   **9.** 300°   **10.** 005°

Section 127
Bearings: scale drawings
Page 155

Exercise 1

1. 760 km   **2.** 910 km   **3.** 5·0 km   **4.** 79 km
5. 37 km   **6.** 98 km   **7.** 26 km   **8.** 560 km

Section 128
Cubes and cuboids
Page 156

Exercise 1

1. Cubes are F, H, J, L, M, O
2. Cuboids are A, B, C, D, E, G, I, K, N.

Exercise 2

3. 12   **4.** 12   **5.** 8   **6.** 8
7. 3   **8.** 3

Section 129
Cubes and nets
Page 157

Exercise 1

1, 2, 6 are nets for a cube.

Section 130
More shapes and nets
Page 158

Exercise 1

2.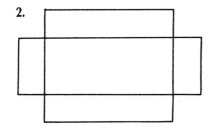

3.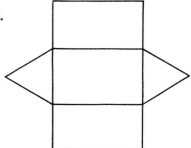

4.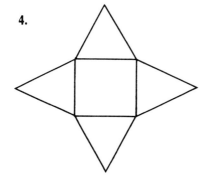

5.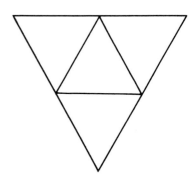

Exercise 2

1.

Shape	Corners	Faces	Edges
cuboid	8	6	12
△ – prism	6	5	9
☐ – based pyramid	5	5	8
tetrahedron	4	4	6

2. Corners + Faces − 2 = Edges

Section 131
Volume
Page 159

Exercise 1

1. 8   **2.** 8   **3.** 16   **4.** 24   **5.** 18   **6.** 21   **7.** 15

Section 132
**Volume
calculations**
Page 160

Exercise 1

1. 10 cm³   **2.** 27 cm³   **3.** 12 cm³   **4.** 20 cm³

Exercise 2

1. 9 cm³   **2.** 9 cm³   **3.** 12 cm³   **4.** 9 cm³   **5.** 18 cm³
6. 8 cm³   **7.** 10 cm³   **8.** 7 cm³   **9.** 3 cm³

Exercise 3

2. 40 mm³   **3.** 40 m³   **4.** 18 km³   **5.** 90 cm³   **6.** 27 cm³

Section 133
Line symmetry
Page 162

Exercise 1

1. 4   **2.** 2   **3.** 2   **4.** 2   **5.** 1   **6.** 0
7. 2   **8.** 3   **9.** 1   **10.** 0   **11.** 5   **12.** 2

Section 135
**Transformations:
reflections**
Page 164

Exercise 1

1.

2.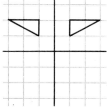

3.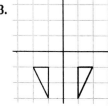

4.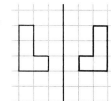

5.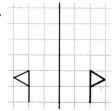

6.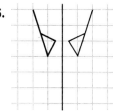

7.

8.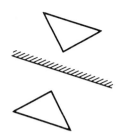

9.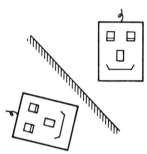

10.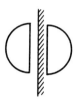

11.

12.

Section 136
Geometry:
Revision
crossword
Page 165

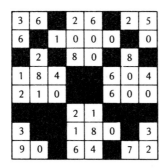

Geometry:
Revision test
Page 171

3a. 270°   **b.** 180°   **c.** 90°

5. 78°

6c. 4 cm, approx.   **d.** Yes, yes

7. Area is 21·5 cm²;   perimeter is 18·6 cm

8. A = 45°   **9.** 43 mm

10a. 5·04 cm²   **b.** 7·92 cm²

11. 490 km, approx.   **12.** 314 cm²

13a. 16 cm²   **b.** 6, 96 cm² **d.** 96 cm²   **e.** the areas are equal

14a. 16   **b.** 10   **c.** 8   **d.** 21

15. 780 000 cm³ = 0·78 m³

Part 4: Graphs and statistics

Section 137
Frequency tables
Page 175

Exercise 1

1a.

marks out of 10	frequency	
	tally	total
0	//	2
1	++++	5
2	++++ /	6
3	++++ ++++	10
4	++++ /	6
5	///	3
6	////	4
7	//	2
8	/	1
9	/	1
10		0

c. The test was hard because most people got less than 5 out of 10.

2a.

marks out of 10	frequency	
	tally	total
0	/	1
1	/ / / /	4
2	++++ ++++ /	11
3	++++ / /	7
4	++++ ++++ / /	12
5	++++ / / /	8
6	++++ / / / /	9
7	++++	5
8	/ / /	3
9		0
10		0

c. This test was hard. Most pupils got less than half marks.

3a. 50 stamps.

b.

country	frequency	
	tally	total
Australia	++++ ++++ / / /	13
Belgium	/ /	2
Canada	++++	5
Denmark	/ / / /	4
Egypt	/ /	2
France	++++ ++++ / / / /	14
Germany	++++ ++++	10

d. France.

4a.

matches	frequency	
	tally	total
46	/ / /	3
47	/ /	2
48	++++ /	6
49	++++ / / / /	9
50	++++ / / /	8
51	++++ /	6
52	++++ / / /	8
53	/ / /	3
54	++++	5

c. 49.

d. Yes. Most of the boxes have contents near fifty.

5a.

shapes	frequency	
	tally	total
Circles	~~//// ~~/ / /	8
Triangles	~~//// ~~/ /	7
Quadrilaterals	~~//// ~~/ / / /	9
Pentagons	~~//// ~~/	6

Most frequent: Quadrilateral
Least frequent: Pentagon

6a.

score	frequency
0–9	1
10–19	2
20–29	10
30–39	5
40–49	9
50–59	6
60–69	6
70–79	1
80–89	3
90–99	2

b. The exam was difficult. Most people got less than half marks.

Section 138
Pie charts
Page 178

Exercise 1

1a. 36 **b.** 10°

c.

distance	frequency	degrees
less than 1 km	10	100
1–2 km	16	160
2–3 km	4	40
over 3 km	6	60

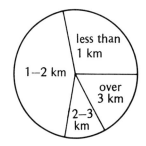

2.

age of people	frequency	degrees
10–19	4	180
20–29	2	90
30 or over	2	90

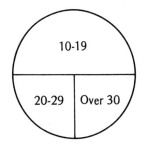

3.

marks	frequency	degrees
2	11	33
3	10	30
4	17	51
5	24	72
6	26	78
7	10	30
8	14	42
9	8	24

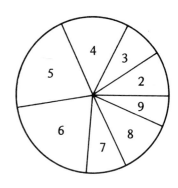

4.

programme	frequency	degrees
Comedy	65	130
Music	67	134
News	9	18
Quiz	23	46
Adverts	16	32

b. 180 minutes; 3 hours
c. 2 degrees represents 1 minute.
d. She likes watching music and comedy programmes.

Section 139
More pie
charts
Page 179

Exercise 1

1a. 2°

b. Cocoa 16   Juice 74   Milk 45   Squash 45

1c.

	Cocoa	Juice	Milk	Squash
No. of drinks sold	16	74	45	45

2a. 3°

b. Badminton 30   5-a-side 32   Swimming 58

2c.

	Badminton	5-a-side	Swimming
No. of participants	30	32	58

Section 140
Pictographs
Page 180

Exercise 1

1a. 10   **b.** 4   **c.** 0
2a. 5   **b.** 2   **c.** Monday   **d.** 16   **e.** £1·12
3a. 100   **b.** 5 cups   **c.** 95
d. No; you do not know the number of teachers in the departments.

Section 141
Bar charts
Page 182

Exercise 1

1a.

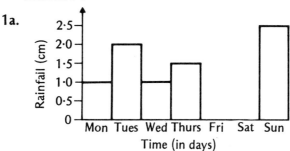

b. Friday and Saturday   **c.** 8 cm

2.

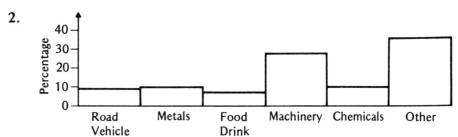

3a. Friday   **b.** Friday   **c.** Saturday
d. Friday or Saturday: Friday is the warmest, but Saturday has the least
change in temperature.
e. Summertime

4.

(1.)

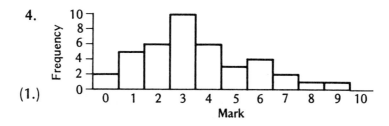

(2.)

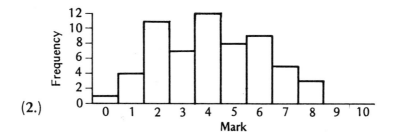

(3.)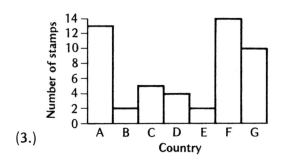

(4.)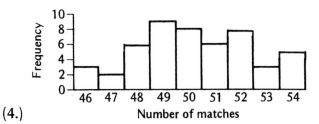

(5.)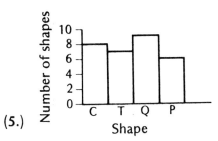

(6.)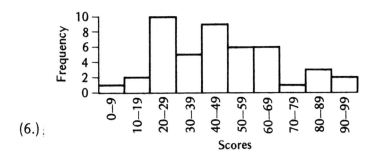

Section 142
Line graphs
Page 184

Exercise 1

1a. 9 cm   **b.** 12 cm   **c.** 13 cm   **d.** 16 cm   **e.** every 2 days

2a. 11°C   **b.** 4°C   **c.** 60 minutes

d. 25 minutes   **e.** Yes. 0°C

f.

0	10	20	30	40	50	60
20	11	6	4	3	2	1

g. 0°C

3a.

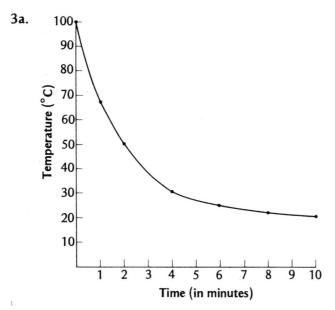

b. Yes – it is reasonable to assume that the water cools less and less quickly as time passes and that this is a continuous process.

c. 20°C   **d.** Yes : 67°C

4a.

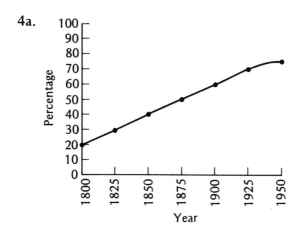

b. Yes – it is reasonable to assume that the population is increased at the same rate between the 'known' times.

c. No, it is not really possible to predict accurately. Anything might happen!   **d.** Yes – about 10% approx.

5a. 3 km   **b.** 10 km   **c.** 10 min   **d.** 8 min
e. between 12th and 20th minutes   **f.** the train had stopped
g. increased the speed of the train

6a.

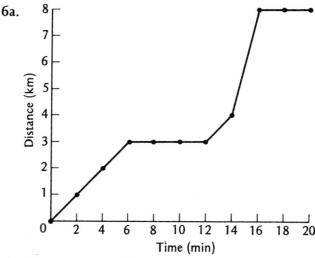

b. 2 km   **c.** 2·5 km — no, not accurately

d. A man drove his car 3 km in 6 minutes then stopped for 6 minutes. He did 1 km in the next 2 minutes and then went much faster, covering 4 km in 2 minutes. He then stopped.

Section 143
Cheating with graphs
Page 187

Exercise 1

1a. £2·20   £2·00   **b.** 20p   **c.** The first one

d. The vertical scales are different   **e.** The first one

2a. The vertical scales are different   **b.** The second one

c. The first one — it looks like the brakes are much more powerful.

3a. The vertical scale is unfair. The line looks to be increasing quickly, but actually the temperature has only risen 1°C.

b. 1°C

c.

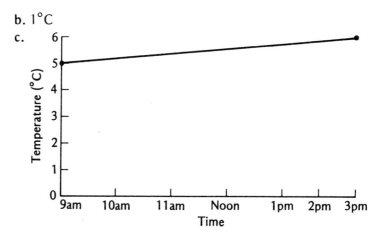

4. No. The drawings have doubled the height as well as the length, making the areas 4 times and 16 times as large, respectively. Money is only two times and 4 times as large, respectively.

5. c	**6.** b	**7.** d	**8.** b	
9. c	**10.** d	**11.** b	**12.** d	

Section 144
**Averages:
the mean**
Page 191

Exercise 1

1a. 7   **b.** 6   **2a.** 16   **b.** 19   **3a.** 46   **b.** 1
4. 42   **5.** 16   **6.** 17   **7.** 16   **8.** 16   **9.** 2412

Section 145
**Mode and
median**
Page 192

Exercise 1

1a. 1   **b.** 4   **2a.** 3   **b.** 5   **3a.** 19   **b.** 130

Exercise 2

1. (1a.) 4   (b.) 4   (2a.) 3   (b.) 2   (3a.) 19   (b.) 130

2a.

number	frequency
0	2
1	1
2	11
3	7
4	6
5	5
6	5
7	3
8	3
9	5

b. 2
c. 4

3a.

heights	frequency
1·80	5
1·81	8
1·82	2
1·83	4
1·84	7
1·85	3

b. 1·82 m
c. 1·81 m

Section 146
**Mean, median
and mode**
Page 193

Exercise 1

1a.

letter	frequency
A	9
B	19
C	17
D	5
E	9

b. B   **c.** C
d. No – you cannot divide letters by a
number.

2a.

score	frequency
0–9	2
10–19	3
20–29	9
30–39	4
40–49	6
50–59	9
60–69	10
70–79	4
80–89	4
90–99	4

b. 60–69   **c.** 50–59

d. Yes — add them all up and divide by 55

3.

position	frequency
1	8
2	7
3	7
4	4
5	1
6	3

Mode = 1st

Section 147
Probability
Page 194

Exercise 1

1. 12   **2.** 2   **3.** 6–1, 1–6, 5–2, 2–5, 4–3, 3–4

4.

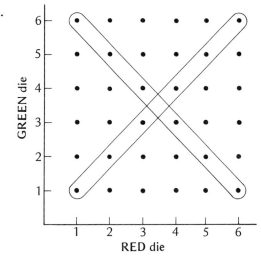

6. No   **7.** 12   **8.** 36

Section 148
**Probability:
single event**
Page 195

Exercise 1

1. $\frac{1}{6}$ **2.** $\frac{1}{13}$ **3.** $\frac{1}{4}$ **4.** $\frac{1}{52}$ **5.** $\frac{1}{2}$ **6.** $\frac{1}{2}$

7. $\frac{4}{51}$ **8.** $\frac{1}{17}$ **9.** $\frac{26}{51}$ **10.** $\frac{4}{7}$ **11.** $\frac{2}{5}$ **12.** $\frac{4}{5}$

Section 149
**Probability:
combinations
of events**
Page 196

Exercise 1

1. $\frac{1}{3}$ **2.** $\frac{1}{2}$ **3.** $\frac{1}{3}$ **4.** $\frac{1}{2}$ **5.** $\frac{26}{51}$

6a. $\frac{13}{17}$ **b.** $\frac{16}{17}$ **c.** 1

7a. $\frac{4}{5}$ **b.** $\frac{3}{5}$ **c.** $\frac{4}{5}$ **d.** 1

Exercise 2

1. $\frac{1}{36}$ 2. $\frac{1}{36}$ 3. $\frac{1}{4}$ 4. $\frac{1}{4}$ 5. $\frac{36}{81} = \frac{4}{9}$

6a. $\frac{1}{10}$ b. $\frac{3}{20}$ c. $\frac{1}{4}$ 7. $\frac{1}{6}$

Graphs and statistics Revision test
Page 198

1a.

size	frequency	degrees
1	5	40
2	6	48
3	10	80
4	12	96
5	7	56
6	5	40

b. 4   c.

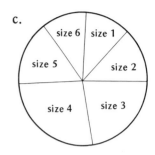

2a. 66   b. 7   c. 26   d. £128·25

3a. 12   b. 20   c. about 1 week

4. A − 4   B − 3   C − 5   D − 1   E − 2

5. 6   6. 61

7a.

cups of tea	frequency
0	3
1	6
2	7
3	6
4	8
5	5
6	5
7	3
8	1
9	1

b.

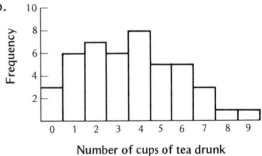

c. 4

8a. $\frac{1}{2}$ b. $\frac{1}{2}$ c. $\frac{1}{2}$ d. $\frac{8}{13}$

9a. $\frac{2}{15}$ b. $\frac{8}{15}$ c. $\frac{7}{15}$ d. $\frac{11}{15}$

10a. $\frac{1}{46\ 656}$

Section 150 **Calculator crossword** *Page 200*

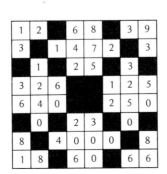